Frederic P. Miller, Agnes F. Vandome,
John McBrewster (Ed.)

Enfield revolver

Frederic P. Miller, Agnes F. Vandome,
John McBrewster (Ed.)

Enfield revolver

Royal Small Arms Factory, Service pistol, Webley Revolver, Smith & Wesson Victory Model, Colt New Service, Smith & Wesson No. 3 Revolver

Alphascript Publishing

Imprint

All parts of this book are extracted from Wikipedia, the free encyclopedia (www.wikipedia.org).

You can get detailed informations about the authors of this collection of articles at the end of this book. The editors (Ed.) of this book are no authors. They have not modified or extended the original texts.

Pictures published in this book can be under different licences than the GNU Free Documentation License. You can get detailed informations about the authors and licences of pictures at the end of this book.

The content of this book was generated collaboratively by volunteers. Please be advised that nothing found here has necessarily been reviewed by people with the expertise required to provide you with complete, accurate or reliable information. Some information in this book maybe misleading or wrong. The Publisher does not guarantee the validity of the information found here. If you need specific advice (f.e. in fields of medical, legal, financial, or risk management questions) please contact a professional who is licensed or knowledgeable in that area.

Cover image: www.ingimage.com
Concerning the licence of the cover image please contact ingimage.

Publisher:
Alphascript Publishing is a trademark of
VDM Publishing House Ltd.,17 Rue Meldrum, Beau Bassin,1713-01 Mauritius
Email: info@vdm-publishing-house.com
Website: www.vdm-publishing-house.com

Published in 2010

Printed in: U.S.A., U.K., Germany. This book was not produced in Mauritius.

ISBN: 978-613-2-84825-3

Contents

Articles

References

Article Licenses

Enfield revolver

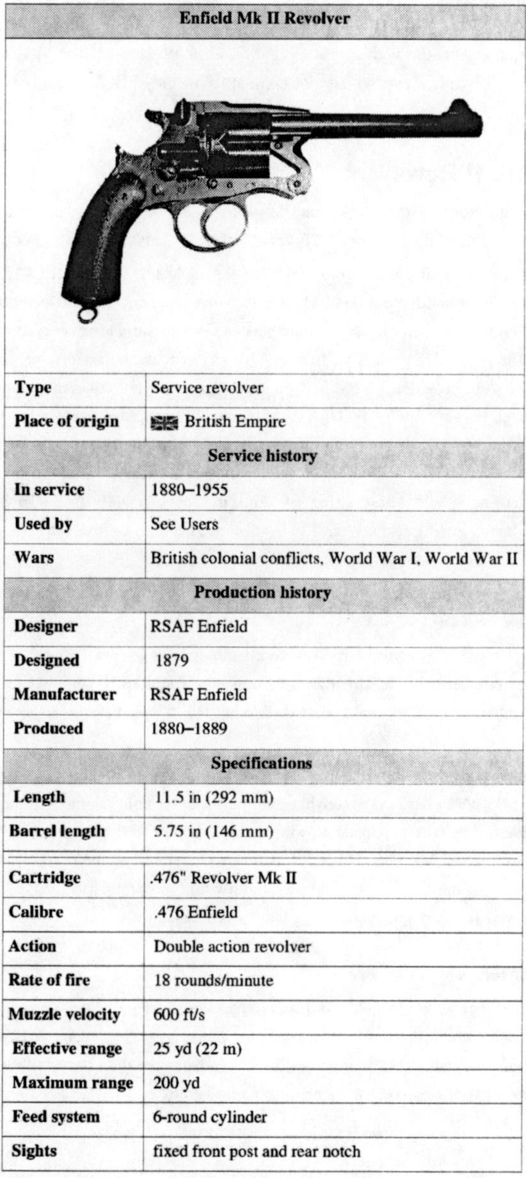

Enfield Mk II Revolver	
Type	Service revolver
Place of origin	British Empire
Service history	
In service	1880–1955
Used by	See Users
Wars	British colonial conflicts, World War I, World War II
Production history	
Designer	RSAF Enfield
Designed	1879
Manufacturer	RSAF Enfield
Produced	1880–1889
Specifications	
Length	11.5 in (292 mm)
Barrel length	5.75 in (146 mm)
Cartridge	.476" Revolver Mk II
Calibre	.476 Enfield
Action	Double action revolver
Rate of fire	18 rounds/minute
Muzzle velocity	600 ft/s
Effective range	25 yd (22 m)
Maximum range	200 yd
Feed system	6-round cylinder
Sights	fixed front post and rear notch

Enfield Revolver is the name applied to two totally separate models of self-extracting British handgun designed and manufactured at the government-owned Royal Small Arms Factory in Enfield; initially the .476 calibre (actually 11.6 mm)[1] Revolver Enfield Mk I/Mk II revolvers (from 1880–1889), and later the .38/200 calibre Enfield No. 2

Mk I (from 1923–1957).

The .476 calibre **Enfield Mk I** and **Mk II** revolvers were the official sidearm of both the British Army and the Northwest Mounted Police—as well as being issued to many other Colonial units throughout the British Empire—and the later model .38/200 **Enfield No. 2 Mk I** revolver was the standard British/Commonwealth sidearm in the Second World War, alongside the Webley Mk IV and Smith & Wesson Victory Model revolvers chambered in the same calibre. The term "Enfield Revolver" is not applied to Webley Mk VI revolvers built by RSAF Enfield between 1923 and 1926.

Enfield Mk I & Mk II Revolvers

The first models of Enfield revolver—the Mark I and Mark II—were official British military sidearms from 1880 through 1887, and issue sidearms of the Northwest Mounted Police in Canada from 1883 until 1911.[2]

NWMP Commissioner Acheson G. Irvine ordered 200 Mark IIs in 1882,[3] priced at C$15.75 each,[4] which were shipped by London's Montgomery and Workman in November that year, arriving in December.[5] They replaced the Adams.[6] Irvine liked them so much, in one of his final acts as Commissioner, he ordered another 600, which were delivered in September 1885;[7] his replacement, Lawrence W. Herchmer, reported the force was entirely outfitted with Enfields (in all 1,079 were provided)[8] and was pleased with them, but concerned about the .476 round being too potent.[9] The first batch was stamped *NWMP-CANADA* (issue number between) after delivery; later purchases were not.[10] They were top-break single- or double-action,[11] and fitted with lanyard rings.[12] Worn spindle arms would fail to hold empty cases on ejection, and worn pivot pins could cause barrels to become loose, resulting in inaccuracy.[13] Its deep rifling would allow firing of slugs of between .449 and .476 in (11.4 and 12.1 mm) diameter.[14] Complaints began arising as early as 1887, influenced in part by the British switching to Webleys,[15] and by 1896, hinge wear and barrel loosening were a real issue.[16]

Beginning in late 1904,[17] the Mark II began to be in favor of the .45 calibre Colt New Service revolver, but the Enfield remained in service until 1911.[18]

The .476 Enfield cartridge the Enfield Mk I/Mk II were chambered for fired a 265 gr (17.2 g) lead bullet, loaded with 18 gr (1.2 g) of black powder.[19] The cartridge was, however, found to be somewhat underpowered during the Afghan War and other contemporary Colonial conflicts, lacking the stopping power believed necessary for military use at the time.

Unlike most other self-extracting revolvers (such as the Webley service revolvers or the Smith & Wesson No. 3 Revolver), the Enfield Mk I/Mk II was somewhat complicated to unload, having an Owen Jones selective extraction/ejection system which was supposed to allow the firer to eject spent cartridges, whilst retaining live rounds in the cylinder. The Enfield Mk I/Mk II had a hinged frame, and when the barrel was unlatched, the cylinder would move forward, operating the extraction system and allowing the spent cartridges to simply fall out. The idea was that the cylinder moved forward far enough to permit fired cases to be completely extracted (and ejected by gravity), but not far enough to permit live cartridges (i.e., those with projectiles still present, and thus longer in overall length) from being removed in the same manner.

The system was obsolete as soon as the Enfield Mk I was introduced, especially as it required reloading one round at a time via a gate in the side (much like the Colt Single Action Army or the Nagant M1895 revolvers). Combined with the somewhat cumbersome nature of the revolver, and a tendency for the action to foul or jam when extracting cartridges, the Enfield Mk I/Mk II revolvers were never popular and eventually replaced in 1889 by the .455 calibre Webley Mk I revolver.

Enfield No. 2 Mk I Revolver

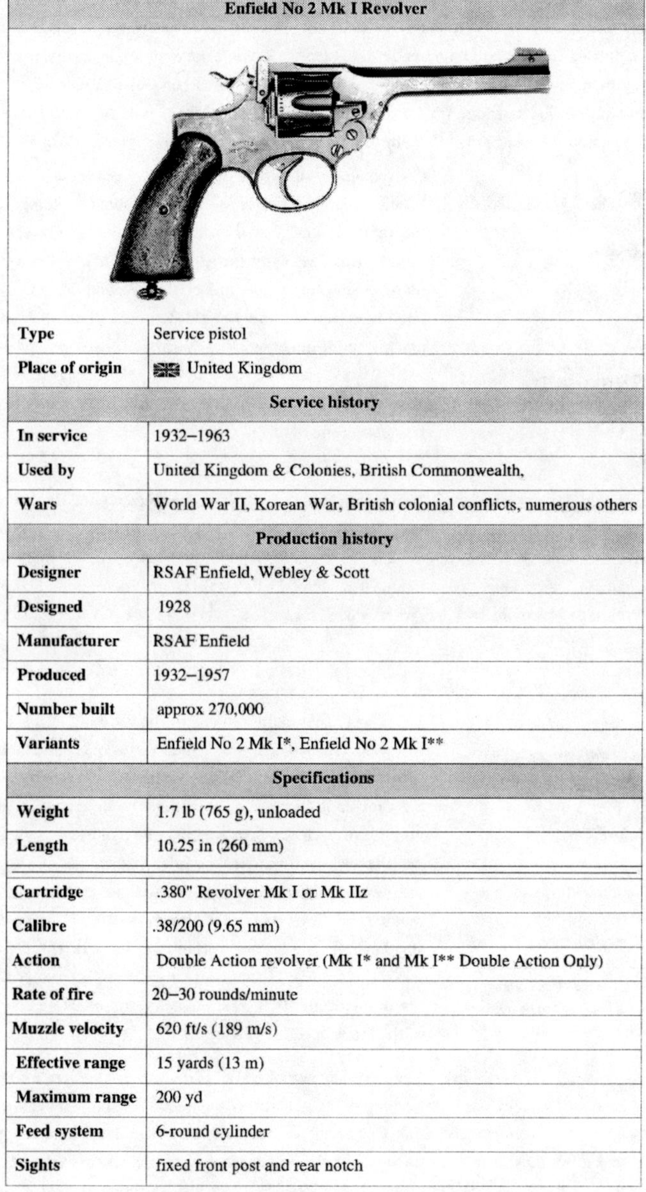

Enfield No 2 Mk I Revolver	
Type	Service pistol
Place of origin	United Kingdom
Service history	
In service	1932–1963
Used by	United Kingdom & Colonies, British Commonwealth,
Wars	World War II, Korean War, British colonial conflicts, numerous others
Production history	
Designer	RSAF Enfield, Webley & Scott
Designed	1928
Manufacturer	RSAF Enfield
Produced	1932–1957
Number built	approx 270,000
Variants	Enfield No 2 Mk I*, Enfield No 2 Mk I**
Specifications	
Weight	1.7 lb (765 g), unloaded
Length	10.25 in (260 mm)
Cartridge	.380" Revolver Mk I or Mk IIz
Calibre	.38/200 (9.65 mm)
Action	Double Action revolver (Mk I* and Mk I** Double Action Only)
Rate of fire	20–30 rounds/minute
Muzzle velocity	620 ft/s (189 m/s)
Effective range	15 yards (13 m)
Maximum range	200 yd
Feed system	6-round cylinder
Sights	fixed front post and rear notch

After the First World War, it was decided by the British Government that a smaller and lighter .38 calibre (9.65 mm) sidearm firing a long, heavy 200 grain (13 g) soft lead bullet would be preferable to the large Webley service revolvers using the .455 calibre (11.6 mm) round.[20] [21] While the .455 had proven to be an effective weapon for stopping enemy soldiers, the recoil of the .455 cartridge complicated marksmanship training.[22] The authorities began a search for a double-action revolver with less weight and recoil that could be quickly mastered by a minimally-trained[23] soldier, with a good probability of hitting an enemy with the first shot at extremely close ranges.[24] By using such a long, heavy, round-nose lead bullet in a .38 calibre cartridge, it was found that the bullet, being minimally stabilised for its weight and calibre, tended to 'keyhole' or tumble longitudinally when striking an object, theoretically increasing wounding and stopping ability of human targets at short ranges.[25] [26] At the time, the .38 calibre Smith & Wesson cartridge with 200-grain (13 g) lead bullet, known as the .38/200, was also a popular cartridge in civilian and police use (in the USA, the .38/200 or 380/200 was known as the *.38 Super Police* load).[27]

Webley Mk IV .38/200 revolver. The similarities between the Webley and Enfield designs are rather obvious.

Consequently, the British firm of Webley & Scott tendered their Webley Mk IV revolver in .38/200 calibre.[28] Rather than adopting it, the British authorities took the design to the Government-run Royal Small Arms Factory at Enfield, and the Enfield factory came up with a revolver that was very similar to the Webley Mk IV .38, but internally slightly different. The Enfield-designed pistol was quickly accepted under the designation *Revolver, No 2 Mk I*, and was adopted in 1931,[29] followed in 1938 by the Mk I* (spurless hammer, double action only),[30] and finally the Mk I** (simplified for wartime production) in 1942.[31]

Webley sued the British Government for £2,250, being "costs involved in the research and design" of the revolver. Their action was contested by Enfield, who stated that the Enfield No 2 Mk I was actually designed by Captain Boys (the Assistant Superintendent of Design, famous for the Boys Rifle) with assistance from Webley & Scott, and not the other way around—accordingly, their claim was denied. By way of compensation, however, the Royal Commission on Awards to Inventors awarded Webley & Scott £1,250.[32]

Variants

Enfield No.2 Mk I* revolver, used by the Tank Corps. The No. 2 Mk I* configuration was double-action only and is therefore missing the hammer spur that would enable it to be manually cocked by the firer.

There were two main variants of the Enfield No 2 Mk I revolver. The first was the **Mk I***, which had a spurless hammer and was double action only, meaning that the hammer could not be thumb-cocked by the shooter for each shot. Additionally, in keeping with the revolver's purpose as a close-range weapon, the handgrips, now made of plastic, were redesigned to improve grip when used in rapid double-action fire; the new handgrip design was given the designation Mk II.[33] The majority of Enfields produced were either Mk I* or modified to that standard.[34] The second variant was the **Mk I****, which was a 1942 variant of the Mk I* simplified in order to increase production, but was discontinued shortly thereafter as a result of safety concerns over some of the introduced modifications.

The vast majority of Enfield No 2 Mk I revolvers were modified to Mk I* during WWII, generally as they came in for repair or general maintenance;[35] the official explanation of the change to the Mk I* version was that the Tank Corps had complained the spur on the hammer was catching on protrusions inside tanks, but most historians nowadays believe that the real reason was that the Mk I* version was cheaper and faster to manufacture.[36] When used in the manner in which British forces trained (rapid double-action fire at very close ranges), the No 2 Mk I* is at least as accurate as any other service pistol of its time, because of the relatively light double action trigger pull. It is not, however, the best choice for deliberately-aimed, long-distance shooting — the double action pull will throw the most competent shooter's aim off enough to noticeably affect accuracy at ranges of more than 15 yards (14 m) or so.[37] Some unit Armourers are known to have retrofitted the Enfield No 2 Mk I* back to the Mk I variant, but this was never an official policy and appears to have been done on an individual basis. Despite officially being declared obsolete at the end of WWII, the Enfield (and Webley revolvers) were not completely phased out in favour of the Browning Hi-Power until April 1969.[38]

The Enfield No 2 is very fast to reload—as are all British top-break revolvers—because of its automatic ejector, which simultaneously removes all six cases from the cylinder.
British combat experience during WWII with the .38/200 Enfield revolvers during WWII seemed to confirm that, *for the average soldier*, the Enfield No. 2 Mk I could be used far more effectively than the bulkier and heavier .455 calibre Webley revolvers that had been issued during WWI.[39] Perhaps because of the relatively long double-action trigger pull compared to other pistols capable of single-action fire[34] , the double-action-only Mk I* revolvers were not popular with troops[34] , many of whom took the first available opportunity to exchange them in favour of Smith & Wesson, Colt, or Webley revolvers.[40]

Ammunition

The Enfield No.2 Mk I was designed for use with the .38 S&W cartridge, now officially termed the 380/200, Revolver Mk I, but also known as the .38/200. It had a 200 gr (13 g). unjacketed round-nose, lead bullet of .359" diameter that developed a muzzle velocity of 620 - 650 ft/s (200 m/s).

Just prior to the outbreak of the Second World War, British authorities became concerned that the soft unjacketed lead bullet used in the 380/200 might be considered as violating the laws of land warfare governing deforming or 'explosive bullets'. A new .38 loading was introduced for use in combat utilizing a 178-grain (11.5 g), gilding-metal jacketed lead bullet; new foresights were issued to compensate for the new cartridge's ballistics and change to the point of aim.[33] The new cartridge was accepted into Commonwealth Service as "Cartridge, Pistol, .380 Mk IIz", firing a 178 - 180 grain (11.7 g) full metal jacket round-nose bullet. The 380/200 Mk I lead bullet cartridge was continued in service, originally restricted to training and marksmanship practice.[33] However, after the outbreak of war, supply exigencies forced British authorities to use both the 380/200 Mk I and the .380 Mk IIz loadings interchangeably in combat. U.S. ammunition manufacturers such as Winchester-Western supplied 380/200 Mk I cartridges to British forces throughout the war.[41]

Other manufacturers

The vast majority of Enfield No 2 revolvers were made by RSAF (Royal Small Arms Factory) Enfield, but wartime necessities meant that numbers were produced elsewhere. Albion Motors in Scotland made the Enfield No 2 Mk I* from 1941–1943, whereupon the contract for production was passed onto Coventry Gauge & Tool Co. By 1945, 24,000[42] Enfield No 2 Mk I* and Mk I** revolvers had been produced by Albion/CG&T. The Howard Auto Cultivator Company (HAC) in New South Wales, Australia tooled up and began manufacturing the Enfield No 2 Mk I* and I** revolvers in 1941, but the production run was very limited (estimated at around 350 or so revolvers in total), and the revolvers produced were criticised for being non-interchangeable, even with other HAC-produced revolvers. Very few HAC revolvers are known to exist, and it is thought by many collectors that most of the HAC revolvers may have been destroyed in the various Australian Gun Amnesties and "Buy-Backs".

Users

- Australia
- Canada
- Commonwealth of Nations
- Gambia[43]
- Lesotho[43]
- United Kingdom

See also

- Colt .38/200 Revolver
- Smith & Wesson .38/200 S&W M&P Revolver
- Webley Mk IV revolver
- .38/200

References

- Barnes, Frank C., ed. by John T. Amber. *Cartridges of the World*, p.175, ".476 Ely/.476 Enfield Mk-3", and p.174, ".455 Revolver MK-1/.455 Colt". Northfield, IL: DBI Books, 1972. ISBN 0-695-80326-3.
- Hogg, Ian V., and John Walter.*Pistols of the World*, 4th Ed. Iola, Wisconsin: Krause Publications, 2004. ISBN 0873494601.
- Maze, Robert J. *Howdah to High Power*. Tucson, Arizona: Excalibur Publications, 2002. ISBN 1-880677-17-2.
- Phillips, Roger F., & Klancher, Donald J. *Arms & [sic] Accoutrements of the Mounted Police 1873-1973*. Bloomfield, ON: Museum Restoration Service, 1982. ISBN 0-919316-84-0.
- Smith, W.H.B. *1943 Basic Manual of Military Small Arms* (facsimile). Harrisburg, Penn.: Stackpole Books, 1979. ISBN 0-8117-1699-6.
- Stamps, Mark, and Ian Skennerton. *.380 Enfield Revolver No 2*. London: Greenhill Books, 1993. ISBN 1-85367-139-8.
- Wilson, Royce. "A Tale of Two Collectables". *Australian Shooter* magazine, March 2006.

External links

- The Corps of the Royal Electrical and Mechanical Engineers Museum of Technology: Pistol Revolver .476 inch Enfield Model 1882 [44]
- The Corps of the Royal Electrical and Mechanical Engineers Museum of Technology: Pistol Revolver .38 inch No 2 Mk I [45] and again [46]

References

[1] Barnes, p.175, ".476 Ely/.476 Enfield Mk-3".

[2] Maze, Robert J: *"Howdah to High Power"*, p. 37. Excalibur Publications, 2002.

[3] Phillips, Roger F., & Klancher, Donald J. *Arms & [sic] Accoutrements of the Mounted Police 1873-1973 (Bloomfield, ON: Museum Restoration Service, 1982), p. 21.*

[4] Phillips & Klancher, p. 207 note 2 to Chapter 3; *Sessional Papers*, Vol. XVIII, No. 1, 1885, p. 164, & *Sessional Papers* 5-7, 1885, p. 265.

[5] Phillips & Klancher, p. 23.

[6] A small number of Adams revolvers remained in possession of RCMP officers until 1888. Phillips & Klancher, p. 207 note 7 to Chapter 3.

[7] Phillips & Klancher, p. 21.

[8] Phillips & Klancher, p. 22.

[9] Phillips & Klancher, p. 22.

[10] Phillips & Klancher, pp. 21 & 23.

[11] Phillips & Klancher, p. 21.

[12] Phillips & Klancher, photo p. 22.
[13] Phillips & Klancher, p. 21.
[14] Phillips & Klancher, p. 21.
[15] Phillips & Klancher, p. 22.
[16] Phillips & Klancher, p. 23.
[17] Phillips & Klancher, p. 23.
[18] Phillips & Klancher, p. 23.
[19] Maze, Robert J: *"Howdah to High Power"* (Excalibur Publications, 2002), p. 32.
[20] Stamps, Mark, and Ian Skennerton, *.380 Enfield Revolver No. 2*, page 9.
[21] Smith, W.H.B, *1943 Basic Manual of Military Small Arms* (facsimile), page 11.
[22] Shore, C. (Capt), *With British Snipers to the Reich*, Paladin Press (1988), pp. 200-201
[23] Weeks, John, *World War II Small Arms*, London: Orbis Publishing Ltd. (1979), p. 76: the standard pistol training ammunition allocation per soldier was only *12 rounds per year*
[24] Shore, C. (Capt), *With British Snipers to the Reich*, Paladin Press (1988), p. 201
[25] Shore, C. (Capt), *With British Snipers to the Reich*, Paladin Press (1988), p. 202
[26] Barnes, Frank C., *Cartridges of the World*, 6th ed. DBI Books (1989), p. 239
[27] Barnes, Frank C., *Cartridges of the World*, 6th ed. DBI Books (1989), p. 239
[28] Maze, Robert J., *Howdah to High Power*, page 103.
[29] § A6862, LoC
[30] § B2289, LoC
[31] § B6712, LoC
[32] Stamps, Mark, and Ian Skennerton, *.380 Enfield Revolver No. 2*, page 12.
[33] Dunlap, Roy, *Ordnance Went Up Front*, Samworth Press (1948), p. 141
[34] Weeks, John, *World War II Small Arms*, London: Orbis Publishing Ltd. (1979), p. 76
[35] § B2289, LoC
[36] Wilson, Royce, "A Tale of Two Collectables", *Australian Shooter* magazine, March 2006.
[37] Smith, W.H.B, *1943 Basic Manual of Military Small Arms* (facsimile), page 11.
[38] Stamps, Mark, and Ian Skennerton, *.380 Enfield Revolver No. 2*, page 118
[39] Smith, W.H.B, *1943 Basic Manual of Military Small Arms* (facsimile), page 11.
[40] Stamps, Mark, and Ian Skennerton, *.380 Enfield Revolver No. 2*, page 79
[41] Shore, C. (Capt.), *With British Snipers to the Reich*, Paladin Press (1989), p. 201
[42] Hogg, Ian V., and John Walter.*Pistols of the World*, 4th Ed.
[43] Hogg, Ian (1989). *Jane's Infantry Weapons 1989-90, 15th Edition*. Jane's Information Group. p. 831. ISBN 0710608896.
[44] http://www.rememuseum.org.uk/arms/pistols/armpr.htm#268
[45] http://www.rememuseum.org.uk/arms/pistols/armpr.htm#291
[46] http://www.rememuseum.org.uk/arms/pistols/armpr.htm#516

Royal Small Arms Factory

Former type	Private, privatized in 1984, - formerly Government Owned
Fate	Dissolved
Successor	Royal Ordnance, then BAE Systems
Founded	1816
Defunct	1988
Headquarters	Enfield Lock, United Kingdom
Products	British military rifles, muskets, swords

The **Royal Small Arms Factory (RSAF)** was a UK government-owned rifle factory in the London Borough of Enfield. The factory produced British military rifles, muskets and swords from 1816. It closed in 1988, but some of its work was transferred to other sites.

RSAF machine shop overlooking basin, November 2007

History

Foundation

The factory was located at Enfield Lock on a marshy island bordered by the River Lea and the River Lee Navigation. It was built on the instructions of the Board of Ordnance near the end of the Napoleonic War. The land was acquired in 1812 and the factory completed by 1816.[1] The site had the advantages of water-power to drive the machinery and the River Lee Navigation for the transportation by barge of raw materials and finished weapons to the River Thames, 15 miles away to be loaded onto sailing ships. Neighbouring farmland was acquired to become a restricted area to test ordnance from the Royal Gunpowder Mill. The RSAF was originally all situated on the east side of the Lea, in the county of Essex in Waltham Abbey parish, Sewardstone hamlet. The course of the river was diverted during the life of the factory, and *part* of the site then fell in Enfield parish. Local boundary changes initiated by SI 1993/1141 after it closed transferred the site entirely from Epping Forest (district) to the London Borough of Enfield.

The original ambitious plans by Captain John By included three mills. Later, the engineer John Rennie recommended the construction of a navigable leat. The leat was made, although only one mill with two waterwheels was completed.

In 1816 the barrel branch was transferred from Lewisham; and by 1818 the lock and finishing branches had been moved to the site, enabling the closure of the Lewisham factory. A sword-making department was set up in 1823.

The Crimean War

The factory fought off the threat of closure in 1831; and remained quite modest in size until the Crimean War of 1853/1856, which resulted in vastly increased production.

By 1856 a machine shop was built on American mass-production lines, using American machinery powered by steam engines. The shop was based on a design by John Anderson and built by the Royal Engineers. The workforce increased to 1000, and by 1860 an average of 1,744 rifles were produced per week.

In 1866 another major expansion took place, when the watermill gave way to steampower. The total number of steam engines grew to sixteen; and by 1887 there were 2,400 employees.

Production of the new model rifle designed by James Paris Lee begun in 1889. The famous Lee Enfield rifle was designed in 1895.

20th century

The factory expanded again in World War I; and in World War II. Two other Royal Ordnance Factories were set up in World War II to manufacture rifles designed at RSAF Enfield, and hence to increase arms output in areas less vulnerable to bombing: ROF Fazakerley and ROF Maltby. Both of these have long been closed.

Decline set in after World War II; and in 1963 half the site was closed.

The Royal Small Arms Factory was privatised in 1984 along with a number of Royal Ordnance Factories to become part of Royal Ordnance Plc; and was later bought by British Aerospace (BAe). They closed the site in 1988.[1]

The significance of RSAF Enfield

The factory was set up because of disappointment with the quality and cost of the existing British weapons used in the Napoleonic War. At this time in Britain, they were built as individual gun components mainly in the Gun Quarter, Birmingham by a number of independent manufacturers and then hand-assembled to produce rifles. These component makers eventually combined to become the Birmingham Small Arms Company. The Enfield factory was intended to improve the quality and to drive down costs.[1]

Weapons designed / built at RSAF Enfield

Almost all the weapons in which the Royal Small Arms Factory had a hand in design or production carry either the word **Enfield** or the letters **EN** in their name;

- Enfield Pattern 1853 Rifle-Musket which used the Minié ball ammunition.
- Snider-Enfield Rifle: an 1866 breechloading version of the 1853 Enfield.
- Enfield revolver: standard issue sidearms, two main versions from 1880 to 1957.
- Martini-Enfield: a conversion of the Martini-Henry rifle to .303 calibre, from 1895.
- Lee-Enfield rifles - using the Lee bolt action. There were 13 variants from 1894 to 1957.

L1A1 rifle

- Pattern 1914 Enfield Rifle: intended as a Lee-Enfield replacement, mainly used by snipers in WWI.
- Bren (Brno + Enfield), .303 Light machine gun from 1935 onwards.
- Sten (Shepherd, Turpin + Enfield) 9mm Sub-machine gun from 1941 to 1953
- Polsten low cost version of 20 mm Oerlikon (acknowledging two Polish designers + Sten (= Shepherd, Turpin + Enfield)), from 1944.
- Taden gun: .280 calibre experimental light machine gun, 1951.
- EM-2: .280 calibre bullpup design experimental assault rifle, 1951.
- L1A1, a British FN FAL version 7.62mm Self Loading Rifle, from 1954.
- TC-10 assault pistol, circa 1970s-80s.
- SA-80 or L85 assault rifle, from 1987.

For weapons manufactured at Enfield before 1853, see British military rifles#Early Enfield rifles

The RSAF, Enfield, was famous for its **Pattern Room** which was a collection, or master set, of every weapon made at RSAF Enfield.[2] After closure this collection was moved to ROF Nottingham; which has since closed. The collection is now held at the Royal Armouries Museum, Leeds.

Closure and reuse of the site

Local government boundary changes meant that the majority of the site was now within the London Borough of Enfield. The necessary outline planning permissions were obtained for site redevelopment; making closure of the site attractive to its new owners.

RSAF Interpretation centre

Closure was announced on 12 August 1987, shortly after privatisation as Royal Ordnance, and the site closed in 1988; the machinery was auctioned off in November 1988. BAe then formed a joint venture with the property company Trafalgar House to redevelop the site.[1]

The majority of the site is now covered by a large housing development called Enfield Island Village.The original machine shop frontage and the older part of the rear structure has been retained and was converted into workshops and retail units by the Enfield Enterprise Agency, making use of European Union (ERDF) funding. The buildings also house the RSAF Interpretation centre which can be viewed by appoinment only.[3]

Community

By 1895, the community had long had its own school (demolished), but it now also had a church (demolished in the 1920s),[4] police station- with three sergeants and nine constables in 1902- and a fire brigade which was manned by one professional and 32 amateurs. Housing conditions in the mid 19th century were poor in the area. The extant Government Row a terrace of cottages was built between two watercourses to house some of the factories workers. Several public houses were opened close to the complex with only the *Greyhound* surviving today (2009) [5] and the brewers Truman & Hanbury became responsible for the catering within the factory [6]

Further reading

- Cherry, Bridget and Pevsner, Nikolaus. *Buildings of England: London 4: North*. Pp 452–3 & 45. ISBN 0-14071049-3.
- Hay, Ian (Maj.Gen. John Hay Beith, CBE, MC) (1949). *R.O.F. The Story of the Royal Ordnance Factories, 1939-1948*. London: His Majesty's Stationery Office.
- Putman, T. and Weinbren, D. (1992). *A Short History of the Royal Small Arms Factory, Enfield*, Centre for Applied Historical Studies, Middlesex University.

External links

- History of Industry in Enfield [7]
- Photos of Enfield Island Village [8]
- The Pattern Room [9]

Geographical coordinates: 51°40′07″N 0°00′58″W

References

[1] Pam, David, (1998). *The Royal Small Arms Factory Enfield & Its Workers*. Enfield: Published by the author. ISBN 0-9532271-0-3.
[2] (N/A) (1973). "Preservation: Royal Small Arms Pattern Room". In: *After The Battle*, **2**, (Pages 42 - 43). ISSN 0306-154-X.
[3] Interpretation centre (http://www.rsaic.org/index.php/80/the-interpretation-centre/) Retrieved 10 June 2008
[4] Photograph of church (http://enfielddeanery.org.uk/lostchurches.htm) Retrieved 14 October 2009
[5] The Greyhound public house (http://www.beerintheevening.com/pubs/s/52/5249/Greyhound/Enfield_Lock) Retrieved 14 October 2009
[6] Godfrey A (notes to) *Old Ordnance Survey Maps: Enfield Lock 1895* Alan Godfrey Maps, ISBN 1841511781 Retrieved 14 October 2009
[7] http://www.enfield.gov.uk/448/Industry%20in%20Enfield%20A%20History.htm
[8] http://www.eivral.com/photos.htm
[9] http://www.smallarmsreview.com/pdf/jan04.pdf

Service pistol

A **service pistol** is any handgun issued to military personnel.

Typically service pistols are revolvers or semi-automatic pistols issued to officers, non-commissioned officers and rear-echelon support personnel for self defense, though service pistols may also be issued to special forces as a backup for their primary weapons. Pistols are not typically issued to front-line infantry.

Before firearms were commonplace, officers typically carried swords instead.

A Soviet political officer, armed with a Tokarev TT-33 service pistol, urges Soviet troops forward against German positions during World War II

History

Prior to the introduction of cartridge-loading firearms, there was little standardisation with regards to the handguns carried by military personnel, although it had been important for officers, artillerymen, and other auxiliary troops to have a means of defending themselves, especially as it was not always practical for them to have a full-length rifle or carbine.

Traditionally, soldiers (infantry & cavalry alike) and officers had carried swords for both personal protection and use in combat. The development of firearms in the mid-14th century changed the way battles were fought, and by the late-15th century it was no longer especially practical to close to hand-to-hand combat range to engage one's opponents, owing to the prevalence of pikes and musket-fire (pike and shot) on the battlefield.

Training was also a factor—it took a very long time to train new recruits in the use of longbows and swords—whereas the basic operation of an arquebus could be taught in a comparatively short time. As a result, swords were retained only by officers (who were less likely to be at the front of the pike-and-musket hedge) and by cavalry, for whom early single-shot handguns were of limited use.

The invention of the revolver in 1836 finally made a *service pistol* practical, as prior to this pistols had largely been single-shot weapons usually of no particular standardized pattern.

Although officers traditionally had been obligated to buy their own weapons, non-commissioned officers (NCOs) and other enlisted personnel were generally issued their weapons (which they were then expected to either pay for or return to the quartermaster if they were promoted). Service pistols, on the other hand, were generally issued to officers, NCO, and others who needed to carry personal weapons as part of their duties. Hence, it was quite common for officers to carry government-issued service pistols in combat.

The first service handguns were revolvers, but the development of Semi-automatic pistols (the first practical example being the Mauser C96 "Broomhandle") gradually led to their replacement by Semi-automatic handguns, such as the well-known German P08 Luger, the first Semi-automatic service pistol to be widely adopted by an industrialised nation. Nowadays, service pistols are almost exclusively self-loading.

The British Army was the last major military service to adopt a Semi-automatic service pistol as a standard sidearm, phasing out their Webley Mk IV, Enfield No 2 Mk I, and Smith & Wesson Victory revolvers in 1969,[1] after which the Browning Hi-Power became the Army's official service pistol.

Modern issue

Special operations soldiers often carry a handgun as a secondary weapon to serve in a supplementary capacity to their primary weapon (a rifle, carbine, submachine gun, or shotgun); this practice is not as prevalent among conventional soldiers. Soldiers who do not serve in a direct combat role are often issued a pistol (such as officers, artillery crews, and other rear-echelon personnel), but conventional riflemen are not generally issued a pistol as part of their standard kit.

The tradition of issuing pistols to officers as a primary weapon is being phased out by many nations. The United States Marine Corps, for example, recently began requiring all enlisted personnel and all officers below the rank of LtCol to carry the M4 Carbine as their primary weapon.

Issue by nation

Angola

Firearm	Type	Calibre	Service
Browning Hi-Power	Semi-automatic	9x19mm Parabellum	1935–present

Argentina

Steyr-Mannlicher M1905 pistol

Firearm	Type	Calibre	Service
Steyr-Mannlicher M1905	Semi-automatic	7.65 mm Mannlicher	1905-19??
M1916/M1927	Semi-automatic	.45 ACP	1916-19??
Ballester-Molina	Semi-automatic	.45 ACP	1938-19??
Browning Hi-Power	Semi-automatic	9mm Parabellum	1935–present

Austria

M1898 Rast & Gasser revolver (in the middle)

Firearm	Type	Calibre	Service
M1870/M1870-74/M1882 Gasser	Revolver	11.25x36R, 11.2x29.5 mm (Montenegrin)	1870-1898
Gasser-Kropatschek M1876	Revolver	9x26R	1876-1898
M1878 Gasser	Revolver	9x26R	1878-1898
Rast-Gasser M1898	Revolver	8mm Rast & Gasser	1898-1945
Roth Steyr M1907	Semi-automatic	8mm Roth Steyr	1907-1945
Steyr M1912	Semi-automatic	9mm Steyr, 9mm Parabellum	1912-1945
Walther P38 / Walther P1	Semi-automatic	9mm Parabellum	1938-1995
Pistol 80	Semi-automatic	9mm Parabellum	1980–present

Bangladesh

Firearm	Type	Calibre	Service
Browning Hi-Power	Semi-automatic	9x19mm Parabellum	1935–present
Type 54 pistol	Semi-automatic	7.62x25mm Tokarev	1970–present[2]
Bangladesh Ordnance Factories Type 92	Semi-automatic	9x19mm Parabellum	2008–present[3]

Belgium

Firearm	Type	Calibre	Service
Nagant M1895	Double-action revolver	7.62x38mmR	1895-1945
Browning Hi-Power	Semi-automatic	9x19mm	1935–present

Bermuda

Firearm	Type	Calibre	Service
Browning Hi-Power	Semi-automatic	9x19mm Parabellum	1935–present

Brazil

Firearm	Type	Calibre	Service
M1911	Semi-automatic	.45 ACP	19??–present
M9 pistol	Semi-automatic	9x19mm Parabellum	19??–present

Canada

Firearm	Type	Calibre	Service
Colt Model 1878	Revolver		1885-1902
Colt New Service	Revolver		1900-1928
Colt Model 1911	Semi-automatic		1914-1945
Smith & Wesson 2nd Model "Hand Ejector"	Revolver		1915-1951
Smith & Wesson "Military & Police"	Revolver		1939-1964
Inglis/Browning High Power	Semi-automatic		1944–present
Browning Hi-Power	Semi-automatic		1935–present
SIG-Sauer P225	Semi-automatic		1991–present
SIG-Sauer P226	Semi-automatic		1991–present

Chile

Firearm	Type	Calibre	Service
Steyr M1912	Semi-automatic		19??
Walther P38	Semi-automatic		19??-
SIG-Sauer P226	Semi-automatic		19xx-

China

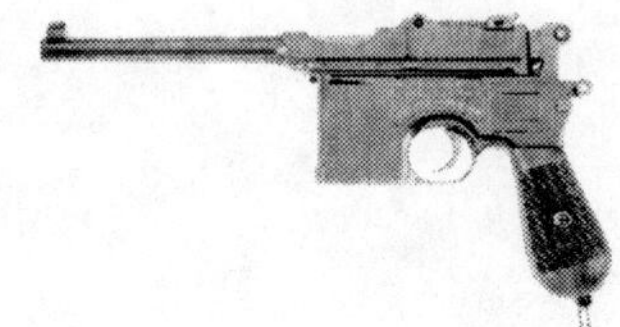

Mauser C96

Firearm	Type	Calibre	Service
Mauser C96 "Broomhandle"	Semi-automatic	7.63x25 Mauser, .45 ACP	1911-1953
Luger P08	Semi-automatic	9mm Parabellum	1911-1949
Browning Hi-Power	Semi-automatic	9mm Parabellum	1935–present
TT pistol	Semi-automatic	7.62x25 Tokarev	1920s–present
Type 51/54	Semi-automatic	7.62x25 Tokarev	1951–present
Type 77	Semi-automatic	7.65x17	1981–present
QSZ-92	Semi-automatic	9mm Parabellum	1996–present

Denmark

Firearm	Type	Calibre	Service
M1861 Danish Navy revolver	Pinfire evolver	11mm Lefaucheux	1861-?
1880 Danish revolver	Double-action revolver		1880-?
M91 Danish Army revolver	Revolver	9mm Danish Army	~1900-~1946
Bergmann-Bayard pistol	Semi-automatic	9x23mm Bergmann	~1910-1946
Neuhausen (Sig P210)	Semi-automatic	9mm Parabellum	~1950–present

Egypt

Firearm	Type	Calibre	Service
Helwan Brigadier	Semi Automatic Pistol	9x19mm Parabellum	

Estonia

Firearm	Type	Calibre	Service
Heckler & Koch USP	Semi-automatic Pistol	9x19mm Parabellum	2007-

Finland

Lahti L-35 service pistol

Firearm	Type	Calibre	Service
Pistol M/19	Semi-automatic	.32 ACP	1919-1971
Pistol M/23	Semi-automatic	7.65mm Parabellum	1922-1980
Lahti L-35	Semi-automatic	9mm Parabellum	1935-1980
Browning Hi-Power	Semi-automatic	9 mm Parabellum	1935–present
9.00 PIST 80-91 (FN HP-DA)	Semi-automatic	9mm Parabellum	1980–present
9.00 PIST 2003 (Walther P99)	Semi-automatic	9mm Parabellum	2003–present

France

Lebel M1892 revolver

Firearm	Type	Calibre	Service
Chamelot Delvigne French 1873	Revolver	11mm	1873-1892
Lebel M1892	Revolver	8mm Lebel	1892-1945
MAB D	Semi-automatic	.32 ACP	1923-1984
Modele 1935	Semi-automatic	7.65x22 Longue	1935-1950
Modele 1950	Semi-automatic	9mm Parabellum	1950-1988
PAMAS	Semi-automatic	9mm Parabellum	1984–present

Germany

Luger P08 service pistol

Firearm	Type	Calibre	Service
M1879 Reichsrevolver	Revolver	10,6 dt. Ordonnanz (10,6x25)	1879-1918
Luger P08	Semi-automatic	9mm Parabellum	1904-1945
Mauser C96 "Broomhandle"	Semi-automatic	7.63x25 Mauser, 9mm Parabellum	1914-1945
Mauser 1914/1934	Semi-automatic	.32 ACP	1914-1945
Walther PP/PPK	Semi-automatic	.32 ACP, .380 ACP, 9mm parabellum	1929–
Walther P38	Semi-automatic	9mm Parabellum	1938–1960s
Mauser HSc	Semi-automatic	.32 ACP	1939-1945
Walther P1	Semi-automatic	9mm Parabellum	1957–present
Heckler & Koch P8	Semi-automatic	9mm Parabellum	1993–present

Hungary

Frommer Stop pistol

Firearm	Type	Calibre	Service
Frommer Stop	Semi-automatic pistol	.32 ACP	1919-1948
M48 Tokarev	Semi-automatic pistol	7.62x25mm	1948-1963
PA-63	Semi-automatic pistol	9mm Makarov	1963-1996
FÉG Model P9RC	Semi-automatic pistol	9mm Parabellum	1996–present

Italy

Beretta M1934 pistol

Firearm	Type	Calibre	Service
Bodeo M1889	Revolver	10.35mm	1889-1945 (?)
Glisenti M1910	Semi-automatic	9mm Glisenti	1910-1945 (?)
Beretta M1923	Semi-automatic	9mm Glisenti	1923-1945 (?)
Beretta M1934	Semi-automatic	.380 ACP	1934-1951
Beretta M1951	Semi-automatic	9mm Parabellum	1951-1981
Beretta 92FS	Semi-automatic	9mm Parabellum	1981–present

Iran

Firearm	Type	Calibre	Service
SIG P226	Semi-automatic	9x19mm Parabellum	19??-

Iraq

Firearm	Type	Calibre	Service
Tariq pistol	Semi-automatic	9x19mm Parabellum	19??-

Japan

Nambu Type 14 pistol

Firearm	Type	Calibre	Service
Meiji Type 26	Revolver	9mm Meiji	1893-1945
Nambu Type 14	Semi-automatic	8mm Nambu	1915-1945
Nambu Type 94	Semi-automatic	8mm Nambu	1934-1945
SIG P220	Semi-automatic	9mm Parabellum	1985–present

Jordan

Firearm	Type	Calibre	Service
Helwan Brigadier	Semi-automatic	9x19mm Parabellum	19??-

Kuwait

Firearm	Type	Calibre	Service
Helwan Brigadier	Semi-automatic	9x19mm Parabellum	19??-

Lebanon

Firearm	Type	Calibre	Service
Helwan Brigadier	Semi-automatic	9x19mm Parabellum	19??-

Libya

Firearm	Type	Calibre	Service
Helwan Brigadier	Semi-automatic	9x19mm Parabellum	19??-

Norway

Firearm	Type	Calibre	Service
Lefaucheux M1864	Revolver	11mm Lefaucheux	1864-1930
Nagant M1883	Revolver	9mm Nagant	1883-1894
Nagant M1893	Revolver	7.5mm Swedish Nagant	1893-1940
Kongsberg M1914	Semi-automatic	.45 ACP	1914-1945
Browning Hi-Power	Semi-automatic	9mm Parabellum	1940-1988
Walther P38	Semi-automatic	9mm Parabellum	1948-1988
Glock P80	Semi-automatic	9mm Parabellum	1988–present

Panama

Firearm	Type	Calibre	Service
Browning Hi-Power	Semi-automatic	9x19mm Parabellum	1935–present

Peru

Firearm	Type	Calibre	Service
JO.LO.AR.	Semi-automatic	.45 ACP	1924-19??
Browning Hi-Power	Semi-automatic	9x19mm Parabellum	19??-

Russia]] / [[Soviet Union

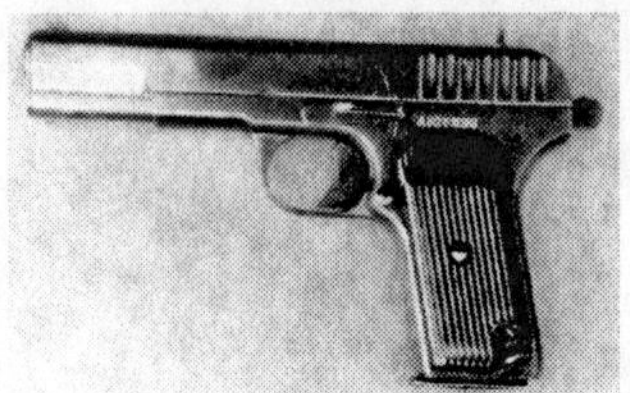

Tokarev TT-33 service pistol

Firearm	Type	Calibre	Service
Smith & Wesson No. 3	Revolver	.44 Russian	1873-1895
Nagant M1895	Revolver	7.62x38R	1895-1950
Mauser C96	Semi-automatic	7.63x25 Mauser	1917-19??
Tokarev TT-33	Semi-automatic	7.62x25 Tokarev	1933–present
Makarov PM	Semi-automatic	9x18 Makarov	1951–present
Stechkin APS	Selective-fire	9x18 Makarov	1951–present
Yarygin PYa	Semi-automatic	9mm Parabellum	2003–present

South Africa

Firearm	Type	Calibre	Service
Browning Hi-Power	Semi-automatic	9x19mm Parabellum	1935–present
Vektor SP1/SP2	Semi-automatic	9x19mm Parabellum	19??-

Sweden

Husqvarna m/1907

Firearm	Type	Calibre	Service
Revolver m/1887	Revolver	7.5 mm Swedish Nagant	1887-1945
Pistol m/07	Semi-automatic	9mm Browning Long	1916-1988
Pistol m/39	Semi-automatic	9mm Parabellum	1939-19??
Pistol m/40	Semi-automatic	9mm Parabellum	1940-1988
Pistol 88	Semi-automatic	9mm Parabellum	1988–present
Pistol 88B	Semi-automatic	9mm Parabellum	1988–present

Switzerland

Model 1882 revolver

Firearm	Type	Calibre	Service
Ordonnanzrevolver 1872	Revolver	10.4mm Swiss rimfire	1872-1878
Ordonnanzrevolver 1878	Revolver	10.4mm Swiss centrefire	1878-1882
Schmidt M1882	Revolver	7.5mm Swiss	1882-1946
Luger pistol	Semi-automatic	7.65 Parabellum	1900-1949
SIG P210	Semi-automatic	9mm Parabellum	1949–present
Walther PPK	Semi-automatic	.32 ACP	1965-19??
SIG P220	Semi-automatic	9mm Parabellum	1975–present

Thailand

Firearm	Type	Calibre	Service
M1911 pistol	Semi-automatic	.45 ACP	19??-

Turkey

Firearm	Type	Calibre	Service
ZIGANA T, C45	Semi-automatic	9x19mm Parabellum, .45 ACP	200?

Tuvalu

Firearm	Type	Calibre	Service
Browning Hi-Power	Semi-automatic	9x19mm Parabellum	1935–present

Ukraine

Firearm	Type	Calibre	Service
Makarov PM	Semi-automatic	9x18mm PM	1951–present
Fort-12	Semi-automatic	9x18mm PM	1997–present
Fort-15	Semi-automatic	9x19mm Parabellum	2006–present

United Arab Emirates

Firearm	Type	Calibre	Service
Browning Hi-Power	Semi-automatic	9mm Parabellum	1971-2007
Caracal F	Semi-automatic	9mm Parabellum	2007–present

United Kingdom, British Empire and Commonwealth of Nations

Webley Mk VI service revolver

A Webley Mark I Revolver, circa 1887, from Canada, cal .455 (Mk I) Webley

Firearm	Type	Calibre	Service
Beaumont-Adams	Revolver	.450 Adams	1853-1880
Enfield Mk I & Mk II	Revolver	.476 Enfield	1880-1887
Webley Mk I-VI	Revolver	.455 Webley	1887-1947
Enfield No 2 Mk I	Revolver	.38/200	1932-1963
Webley Mk IV	Revolver	.38/200	1932-1963
Smith & Wesson M&P or Victory Model	Revolver	.38/200	1940-1963
Browning Hi-Power	Semi-automatic	9mm Parabellum	1940–present

SIG P226	Semi-automatic	9mm Parabellum	1995-present

United States

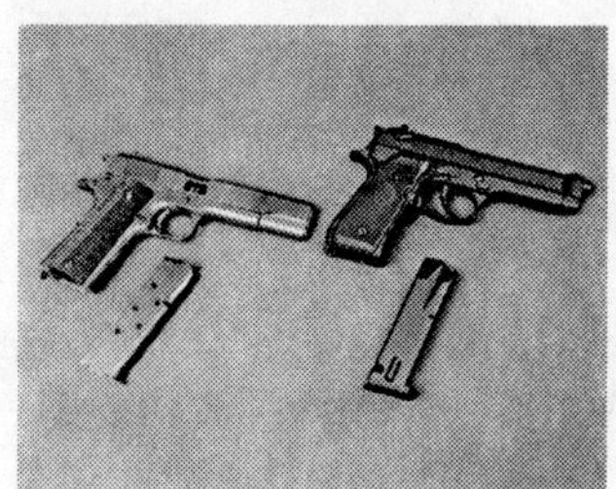

Colt M1911A1 (left) and Beretta M9 (right) service pistols

Firearm	Type	Calibre	Service
Harpers Ferry Model 1805	Flintlock	.58 calibre	1805-?
Colt M1851 Navy	Cap and ball revolver, single-action	.36 Ball	1851-1873
Colt Army Model 1860	Cap and ball revolver, single-action	.44 Ball	1860-1873
Remington Model 1858	Cap and ball revolver, single-action	.36, .44 Ball	1862-1875
Colt Single Action Army	Single-action cartridge revolver	.45 Long Colt	1873-1892
Colt M1892	Double-action revolver	.38 Long Colt	1892-1911
M1911	Semi-automatic	.45 ACP	1911–present
M1917	Revolver	.45 ACP	1917-1953
Smith & Wesson Model 10	Revolver	.38 Special	1935-1972
Browning HP	Semi-automatic	9mm Parabellum	1935–present
M9 pistol	Semi-automatic	9mm Parabellum	1985–present
M11 pistol	Semi-automatic	9mm Parabellum	1985–present
SIG P229 DAK[4]	Semi-automatic	.40 S&W	2006–present

Vietnam

Firearm	Type	Calibre	Service
Makarov PM	Semi-automatic	9x18mm	1951–present

See also

- Service rifle

References

- *Howdah To High Power* (2002) Maze, Robert J, Excalibur Publications, Tucson AZ (USA) ISBN 1-880677-17-2
- *Small Arms Identification Series No. 9: .455 Pistol, Revolver No 1 Mk VI* (1997) Skennerton, Ian, Arms & Militaria Press, Gold Coast QLD (Australia), ISBN 0-949749-30-3
- *.380 Enfield Revolver No 2* (1993) Stamps, Mark and Skennerton, Ian, Greenhill Books, London (UK) ISBN 1-85367-139-8
- *1942 Basic Manual of Military Small Arms (Facsimile Edition)*, Smith, W.H.B, Stackpole Books, Harrisburg PA (USA), ISBN 0-8117-1699-6

References

[1] Stamps, Mark, and Ian Skennerton, *.380 Enfield Revolver No. 2*, page 118

[2] Type 54 pistol on Bangladesh Military Forces website. (http://www.bdmilitary.com/index.php?option=com_content&view=article&id=186&Itemid=95') Retrieved on April 16, 2010.

[3] Type 92 pistol on Bangladesh Military Forces website. (http://www.bdmilitary.com/index.php?option=com_content&view=article&id=99&Itemid=95') Retrieved on April 16, 2010.

[4] Coast Guard only; replaced M9

Webley Revolver

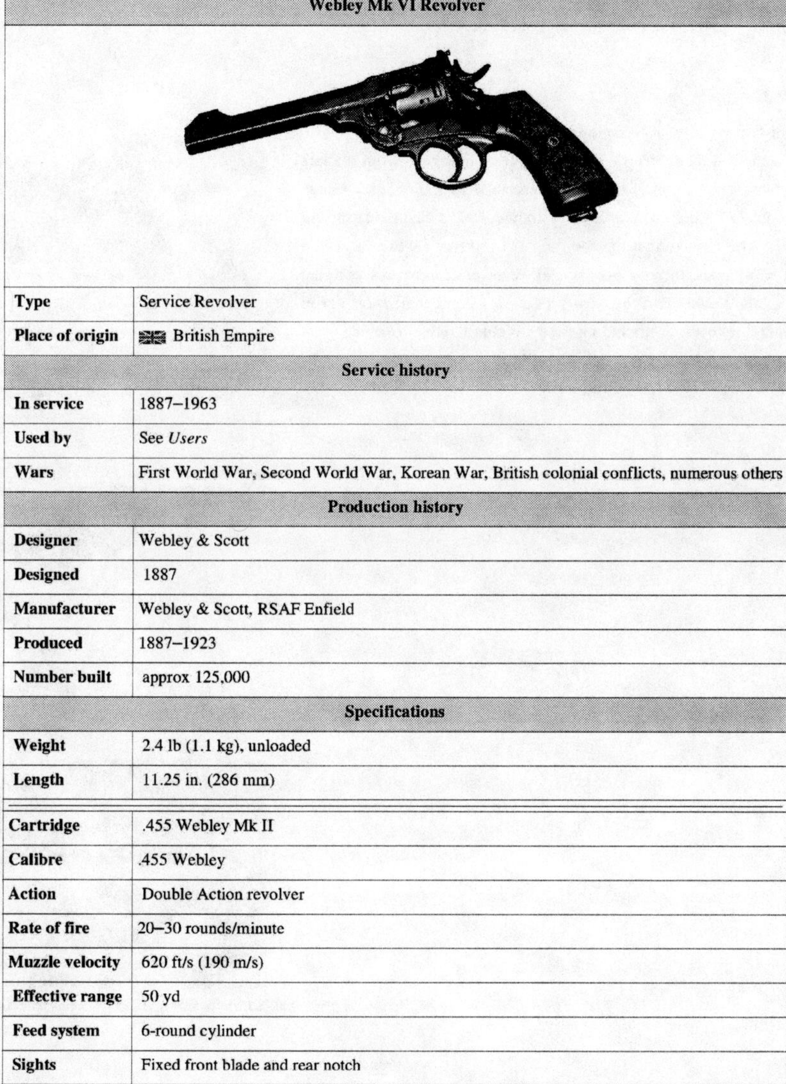

Webley Mk VI Revolver	
Type	Service Revolver
Place of origin	British Empire
Service history	
In service	1887–1963
Used by	See *Users*
Wars	First World War, Second World War, Korean War, British colonial conflicts, numerous others
Production history	
Designer	Webley & Scott
Designed	1887
Manufacturer	Webley & Scott, RSAF Enfield
Produced	1887–1923
Number built	approx 125,000
Specifications	
Weight	2.4 lb (1.1 kg), unloaded
Length	11.25 in. (286 mm)
Cartridge	.455 Webley Mk II
Calibre	.455 Webley
Action	Double Action revolver
Rate of fire	20–30 rounds/minute
Muzzle velocity	620 ft/s (190 m/s)
Effective range	50 yd
Feed system	6-round cylinder
Sights	Fixed front blade and rear notch

The **Webley Revolver** (also known as the **Webley Break-Top Revolver** or **Webley Self-Extracting Revolver)** was, in various marks, the standard issue service pistol for the armed forces of the United Kingdom, the British Empire, and the Commonwealth from 1887 until 1963.

The Webley is a top-break revolver with automatic extraction; breaking the revolver open for reloading also operates the extractor, removing the spent cartridges from the cylinder. The **Webley Mk I** service revolver was adopted in

1887, but it was a later version, the **Mk IV**, which rose to prominence during the Boer War of 1899–1902. The **Mk VI**, introduced in 1915 during the First World War, is perhaps the best-known model.

Webley service revolvers are among the most powerful top-break revolvers ever produced, firing the .455 Webley cartridge. Although the .455 calibre Webley is no longer in military service, the .38/200 Webley Mk IV variant is still in use as a police sidearm in a number of countries.[1]

History

The British company Webley and Scott (P. Webley & Son before merger with W & C Scott) produced a range of revolvers from the late 19th to late 20th centuries. Early models such as the Webley-Green army model 1879 and the Webley-Pryse model were first made during the 1870s. The best-known are the range of military revolvers, which were in service use across two World Wars and numerous colonial conflicts, but Webley & Scott also produced a number of short-barrel solid-frame revolvers, including the Webley RIC (Royal Irish Constabulary) model and the British Bulldog revolver, designed to be carried in a coat pocket for self-defence.

A Webley Mark I Revolver, circa 1887, from Canada, cal .455 (Mk I) Webley

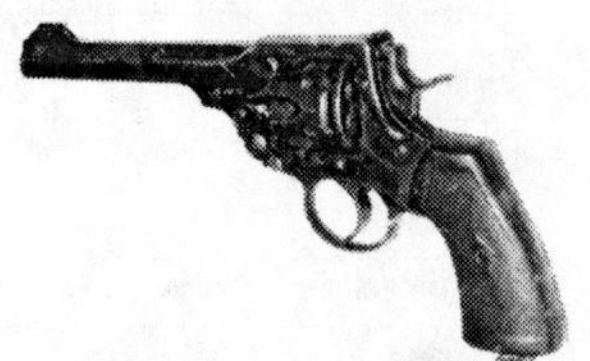

Webley Mark VI .455 service revolver

Close up of the cylinder (including thumb catch) on a Webley Mk VI service revolver

In 1887, the British Army was searching for a revolver to replace the largely unsatisfactory Enfield Mk I & Mk II Revolvers, and Webley & Scott, who were already very well known makers of quality guns and had sold many pistols on a commercial basis to military officers and civilians alike, tendered the .455 calibre Webley Self-Extracting Revolver for trials. The military was suitably impressed with the revolver (it was seen as a vast improvement over the Enfield revolvers then in service, which lacked a practical extraction system), and it was adopted on 8 November 1887 as the "Pistol, Webley, Mk I".[2] The initial contract called for 10,000 Webley revolvers, at a price of £3/1/1- each, with at least 2,000 revolvers to be supplied within eight months.[3]

Webley Royal Irish Constabulary Revolver Cal 450 CF

The Webley revolver went through a number of changes, culminating in the Mk VI, which was in production between 1915 and 1923, finally being retired in 1947, although the Webley Mk IV .38/200 remained in service until 1963 alongside the Enfield No. 2 Mk I revolver. Commercial versions of all Webley service revolvers were also sold to the civilian market, along with a number of similar designs (such as the **Webley-Government** and **Webley-Wilkinson**) that were not officially adopted for service, but were nonetheless purchased privately by military officers.

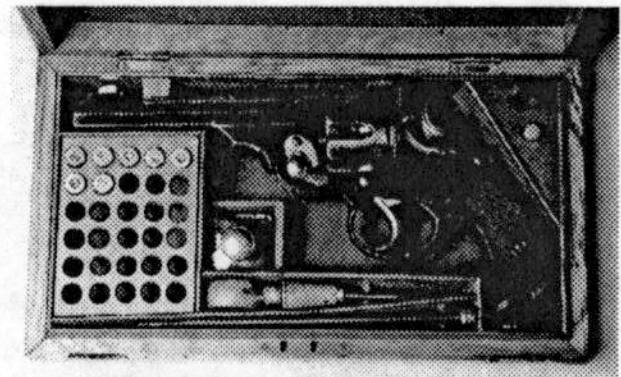
Webley WG (a.k.a. Webley Green) Revolver cal 455/476 (.476 Enfield)

Webley revolvers in military service

.455in SAA Ball ammunition.

Boer War

The Webley Mk IV, chambered in .455 Webley, was introduced in 1899 and soon became known as the "Boer War Model",[4] on account of the large numbers of officers and Non-commissioned officers who purchased it on their way to take part in the conflict. The Webley Mk IV served alongside a large number of other handguns, including the Mauser C96 "Broomhandle" (as used by Winston Churchill during the War), earlier Beaumont-Adams cartridge revolvers, and other top-break revolvers manufactured by gunmakers such as William Tranter, and Kynoch.

First World War

The standard-issue Webley revolver at the outbreak of the First World War was the Webley Mk V (adopted 9 December 1913[5]), but there were considerably more Mk IV revolvers in service in 1914,[6] as the initial order for 20,000 Mk V revolvers had not been completed when hostilities began.[7] On 24 May 1915, the Webley Mk VI was adopted as the standard sidearm for British and Commonwealth troops[7] and remained so for the duration of the First World War, being issued to officers, airmen, naval crews, boarding parties, trench raiders, machine-gun teams, and tank crews. The Mk VI proved to be a very reliable and hardy weapon, well suited to the mud and adverse conditions of trench warfare, and several accessories were developed for the Mk VI, including a bayonet (made from a converted French Gras bayonet),[8] a speedloader device ("Prideaux Device"),[9] and a stock allowing for the revolver to be converted into a carbine.[10]

Second World War

The official service pistol for the British military during the Second World War was the Enfield No. 2 Mk I .38/200 calibre revolver,[11] but owing to a critical shortage of handguns, a number of other weapons were also adopted (first practically, then officially) to alleviate the shortage. As a result, both the Webley Mk IV in .38/200 and the .455 calibre Webley Mk VI were issued to personnel during the war.[12]

A box of Second World War dated .380" Revolver Mk IIz cartridges

Post-war

The Webley Mk VI (.455) and Mk IV (.38/200) revolvers were still issued to British and Commonwealth Forces after the Second World War; there were now extensive stockpiles of the revolvers in military stores. An armourer stationed in West Germany recalled (admittedly tongue-in-cheek) that by the time they were officially retired in 1963, the ammunition allowance was "two cartridges per man, per year." This lack of ammunition was instrumental in keeping the Enfield and Webley revolvers in use so long: they were not wearing out because they were not being used.[13]

The Webley Mk IV .38 revolver was not completely replaced by the Browning Hi-Power until 1963, and saw combat in the Korean War, the Suez Crisis, Malayan Emergency, and the Rhodesian Bush War. Many Enfield No. 2 Mk I revolvers were still floating about in British Military service as late as 1970.[14]

Police use

The (Royal) Hong Kong Police and Royal Singaporean Police were issued Webley Mk III & Mk IV .38/200 revolvers from the 1930s. Singaporean police Webleys were equipped with safety catches, a rather unusual feature in a revolver. These were gradually retired in the 1970s as they came in for repair, and were replaced with Smith & Wesson Model 10 .38 revolvers. The London Metropolitan Police were also known to use Webley revolvers, as were most colonial police units until just after the Second World War. There may still be some police units with Webley Mk IV revolvers that, whilst not issued, are still present in the armoury.

IOF .32 Revolver.

The Ordnance Factory Board of India still manufactures .380 Revolver Mk IIz cartridges,[15] as well as a .32 caliber revolver (also known as IOF Mk1) with 2-inch (51 mm) barrel that is clearly based on the Webley Mk IV .38 service pistol.[16]

Military service .455 Webley revolver marks and models

There were six different marks of .455 calibre Webley British Government Model revolvers approved for British military service at various times between 1887 and the end of the First World War:

- **Mk I**: The first Webley self-extracting revolver adopted for service, officially adopted 8 November 1887, with a 4-inch (100 mm) barrel and "bird's beak" style grips. **Mk I*** was a factory upgrade of Mk I revolvers to match the Mk II.
- **Mk II**: Similar to the Mk I, with modifications to the hammer and grip shape, as well as a hardened steel shield for the blast-shield. Officially adopted 21 May 1895, with a 4-inch (100 mm) barrel.[17]
- **Mk III**: Identical to Mk II, but with modifications to the cylinder cam and related parts. Officially adopted 5 October 1897, but never issued.[18]
- **Mk IV**: The "Boer War" Model. Manufactured using much higher quality steel and case hardened parts, with the cylinder axis being a fixed part of the barrel and modifications to various other parts, including a re-designed blast-shield. Officially adopted 21 July 1899, with a 4-inch (100 mm) barrel.[19]
- **Mk V**: Similar to the Mk IV, but with cylinders 0.12-inch (3.0 mm) wider to allow for the use of nitrocellulose propellant-based cartridges. Officially adopted 9 December 1913, with a 4-inch (100 mm) barrel, although some models produced in 1915 had 5-inch (130 mm) and 6-inch (150 mm) barrels.[20]
- **Mk VI**: Similar to the Mk V, but with a squared-off "target" style grip (as opposed to the "bird's-beak" style found on earlier marks and models) and a 6-inch (150 mm) barrel. Officially adopted 24 May 1915,[21] and also manufactured by RSAF Enfield under the designation **Pistol, Revolver, Webley, No. 1 Mk VI** from 1921–1926.[22]

The Webley Mk IV .38/200 Service Revolver

Webley Mk IV .38/200 Service Revolver	
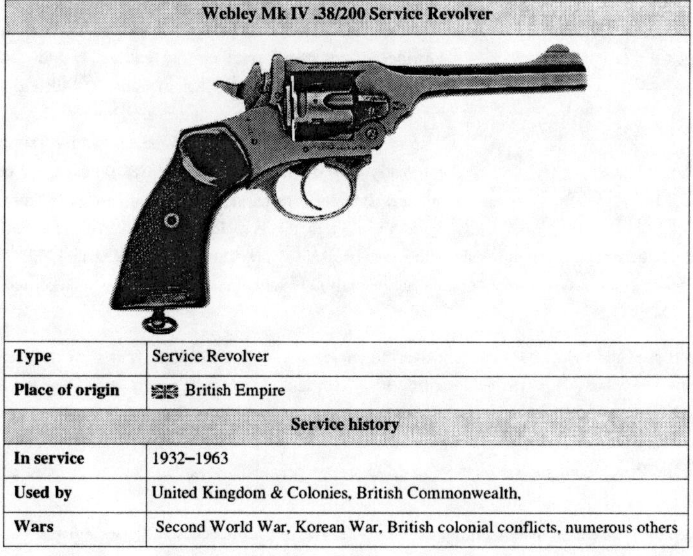	
Type	Service Revolver
Place of origin	British Empire
Service history	
In service	1932–1963
Used by	United Kingdom & Colonies, British Commonwealth,
Wars	Second World War, Korean War, British colonial conflicts, numerous others

Production history	
Designer	Webley & Scott
Designed	1932
Manufacturer	Webley & Scott
Produced	1932–1978
Number built	approx 500,000
Specifications	
Weight	2.4 lb (1.1 kg), unloaded
Length	10.25in. (266 mm)
Cartridge	.380" Revolver Mk IIz
Calibre	.38/200
Action	Double Action revolver
Rate of fire	20–30 rounds/minute
Muzzle velocity	620 ft/s (190 m/s)
Effective range	50 yd
Maximum range	300 yd
Feed system	6-round cylinder
Sights	fixed front post and rear notch

At the end of the First World War, the British military decided that the .455 calibre gun and cartridge was too large for modern military use, and decided (after numerous tests and extensive trials) that a pistol in .38 calibre , firing a 200-grain (13 g) bullet, would be just as effective as the .455 for stopping an enemy.[23]

An Enfield No. 2 Mk I revolver—clearly based on the Webley revolver, if not an outright copy.

Webley & Scott immediately tendered the .38/200 calibre Webley Mk IV revolver, which as well as being nearly identical in appearance to the .455 calibre Mk VI revolver (albeit scaled down for the smaller cartridge), was based on their .38 calibre Webley Mk III pistol, designed for the police and civilian markets.[24] Much to their surprise, the British Government took the design to the Royal Small Arms Factory at Enfield Lock, which came up with a revolver that was externally very similar looking to the .38/200 calibre Webley Mk IV , but was internally different enough that no parts from the Webley could be used in the Enfield and vice-versa. The Enfield-designed pistol was quickly accepted under the designation *Revolver, No. 2 Mk I*, and was adopted in 1932,[25] followed in 1938 by the Mk I* (spurless hammer, double action only),[26] and finally the Mk I** (simplified for wartime production) in 1942.[27]

Webley & Scott sued the British Government over the incident, claiming £2250 as "costs involved in the research and design" of the revolver. This was contested by RSAF Enfield, which quite firmly stated that the Enfield No. 2 Mk I was designed by Captain Boys (the Assistant Superintendent of Design, later of Boys Anti-Tank Rifle fame) with assistance from Webley & Scott, and not the other way around. Accordingly, their claim was denied. By way of compensation, the Royal Commission on Awards to Inventors eventually awarded Webley & Scott £1250 for their work.[28]

RSAF Enfield proved unable to manufacture enough No. 2 revolvers to meet the military's wartime demands, and as a result Webley's Mk IV was also adopted as a standard sidearm for the British Army.

Other well-known Webley Revolvers

Whilst the top-break, self-extracting revolvers used by the British and Commonwealth militaries are the best-known examples of Webley Revolvers, the company produced a number of other highly popular revolvers largely intended for the police and civilian markets.

Webley RIC

The Webley RIC (Royal Irish Constabulary) model was Webley's first double-action revolver, and adopted by RIC in 1868,[29] hence the name. It was a solid frame, gate-loaded revolver, chambered in .442 Webley. General George Armstrong Custer was known to have owned a pair, which he used at the Battle of the Little Bighorn in 1876.[30] [31]

British Bulldog

The British Bulldog series of revolvers were an enormously successful solid-frame design featuring a 2.5-inch (64 mm) barrel and chambered in a variety of heavy-duty calibres, including .442 Webley and .450 Adams. They were designed to be carried in a coat pocket or kept on a night-stand, and great numbers have survived to the present day in good condition, having seen little actual use.[32] Numerous copies of this design were made in France and Belgium (primarily the latter) during the late 19th and early 20th centuries,[33] and they remained reasonably popular until the Second World War. They are now generally sought after as collector's pieces, especially as ammunition for them is (for the most part) no longer commercially available.

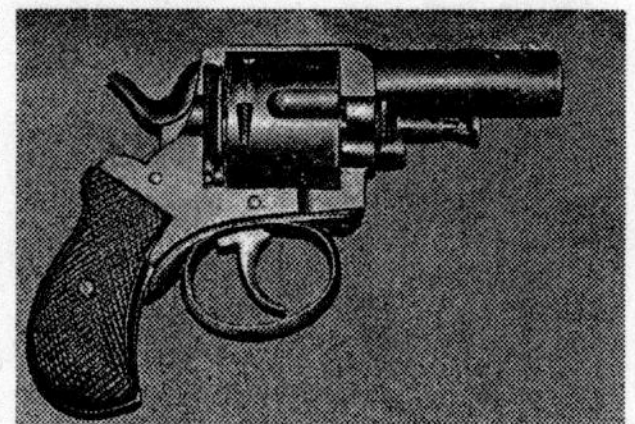

A non-firing replica of a Webley "British Bulldog" revolver.

Webley-Fosbery Automatic Revolver

A highly unusual example of an "automatic revolver", the Webley-Fosbery Automatic Revolver was produced between 1900 and 1915, and available in both a six-shot .455 Webley version, and an eight-shot .38 ACP (not to be confused with .380 ACP) version.[34] Unusually for a revolver, the Webley-Fosbery had a safety catch, and the light trigger pull, solid design, and reputation for accuracy ensured that the Webley-Fosbery remained popular with target shooters long after production had finished.[35] [36]

Cultural impact

Webley Revolvers often serve as a stereotypical British revolver in film and television—their appearance in the film *Zulu*, for example, is an anachronism, as the film is set in 1879 and the Webley Mk VI revolvers shown in use by the British officers were not introduced until 1915, but the Mk VI is based on designs from around the period in which the film is set, and can thus be seen as a stand-in for the historically correct (but more difficult to obtain) Beaumont-Adams Revolver.

Copy of Webley Pocket Pistol in .38 S&W, at Bagram Airfield, Afghanistan

Users

- Commonwealth of Nations[37]
- United Kingdom[37]

References

- Dowell, William Chipchase, *The Webley Story*, Commonwealth Heritage Foundation, Kirkland, WA (USA), 1987. ISBN 0-939683-04-0.
- H.M. Stationer's Office, *List of Changes in British War Material*, H.M.S.O, London (UK), Periodical.
- Maze, Robert J., *Howdah to High Power: A Century of Breechloading Service Pistols (1867–1967)*, Excalibur Publications, Tucson, AZ (USA), 2002. ISBN 1-880677-17-2.
- Skennerton, Ian D., *Small Arms Identification Series No. 9: .455 Pistol, Revolver No. 1 Mk VI*, Arms & Militaria Press, Gold Coast, QLD (Australia), 1997. ISBN 0-949749-30-3.
- Smith, W. H. B., *1943 Basic Manual of Military Small Arms* (Facsimile), Stackpole Books, Harrisburg, PA (USA), 1979. ISBN 0-8117-1699-6.
- Stamps, Mark & Skennerton, Ian D., *.380 Enfield Revolver No. 2*, Greenhill Books, London (UK), 1993. ISBN 1-85367-139-8.
- Wilson, Royce, *A Tale of Two Collectables*, *Australian Shooter* Magazine, March 2006.
- Gerard Henrotin, *Webley Service Revolvers*, H&L Publishing (Belgium), 2007

External links

- The Weapons Collection: Pistols Revolver [44], REME Museum of Technology.
- Historical Overview of the Webley Mk 4 Revolver from *American Rifleman* [38]

References

[1] Historic firearm of the month, July 1999 (http://www.cruffler.com/historic-july99.html), Cruffler.com. Retrieved on 2006-12-02

[2] § 6075, *List of Changes in British War Material* (hereafter referred to as "LoC"), H.M. Stationer's Office, periodical

[3] Skennerton, Ian D., *Small Arms Identification Series No. 9: .455 Pistol, Revolver No. 1 Mk VI*, p. 6, Arms & Militaria Press, 1997.

[4] Maze, Robert J., *Howdah to High Power*, p. 44, Excalibur Publications, 2002.

[5] Dowell, William Chipchase, *The Webley Story*, p. 115. Commonwealth Heritage Foundation, 1987.

[6] Dowell, William Chipchase, *The Webley Story*, p. 114, Commonwealth Heritage Foundation, 1987.

[7] Dowell, William Chipchase, *The Webley Story*, p. 115, Commonwealth Heritage Foundation, 1987.

[8] Dowell, William Chipchase, *The Webley Story*, p. 116, Commonwealth Heritage Foundation, 1987.

[9] Dowell, William Chipchase, *The Webley Story*, p. 178, Commonwealth Heritage Foundation, 1987.

[10] Maze, Robert J., *Howdah to High Power*, p. 49, Excalibur Publications, 2002.

[11] Smith, W. H. B., *1943 Basic Manual of Military Small Arms* (Facsimile), p. 11, Stackpole Books, 1979.

[12] Stamps, Mark & Skennerton, Ian D., *.380 Enfield Revolver No. 2*, p. 87, Greenhill Books, 1993.

[13] Stamps, Mark & Skennerton, Ian D., *.380 Enfield Revolver No. 2*, p. 117, Greenhill Books, 1993.

[14] Stamps, Mark & Skennerton, Ian D., *.380 Enfield Revolver No. 2*, p. 119, Greenhill Books, 1993.

[15] "Cartridge SA .380" Ball Revolver" (http://ofbindia.gov.in/products/data/ammunition/sc/10.htm). Indian Ordnance Factories. . Retrieved 2006-08-03.

[16] "Revolver 32 (7.65 mm x 23)" (http://ofbindia.gov.in/products/data/weapons/wsc/3.htm). Indian Ordnance Factories. . Retrieved 2006-08-03.

[17] § 7816, LoC

[18] § 9039, LoC

[19] § 9787, LoC

[20] § 16783, LoC

[21] § 17319, LoC

[22] Skennerton, Ian D., *Small Arms Identification Series No. 9: .455 Pistol, Revolver No. 1 Mk VI*, p. 10, Arms & Militaria Press, 1997.

[23] Stamps, Mark & Skennerton, Ian D., *.380 Enfield Revolver No. 2*, p. 9, Greenhill Books, 1993; Smith, W. H. B., *1943 Basic Manual of Military Small Arms* (Facsimile), p. 11, Stackpole Books, 1979.

[24] Maze, Robert J., *Howdah to High Power*, p. 103, Excalibur Publications, 2002.

[25] § A6862, LoC

[26] § B2289, LoC

[27] § B6712, LoC

[28] Stamps, Mark & Skennerton, Ian D., *.380 Enfield Revolver No. 2*, p. 12, Greenhill Books, 1993.

[29] Maze, Robert J., *Howdah to High Power*, p. 30, Excalibur Publications, 2002.

[30] Doerner, John A. "Lt. Col. George Armstrong Custer at the Battle of the Little Bighorn" (http://www.geocities.com/burntumber60/MPdoerner.html). Martin Pate. . Retrieved 2006-08-03.

[31] Gallear, Mark (2001). "Guns at the Little Bighorn" (http://www.westernerspublications.ltd.uk/CAGB Guns at the LBH.htm). Custer Association of Great Britain. . Retrieved 2006-08-03.

[32] Ficken, H. R.. "Webley's The British Bull Dog Revolver, Serial Numbering and Variations" (http://members.aol.com/hrftx/TBBD.htm). . Retrieved 2006-08-03.

[33] Kekkonen, P.T.. "British Bulldog revolver" (http://guns.connect.fi/gow/QA14.html). Gunwriters. . Retrieved 2006-08-03.

[34] Dowell, William Chipchase, *The Webley Story*, p. 128, Commonwealth Heritage Foundation, 1987.

[35] Maze, Robert J., *Howdah to High Power*, p. 78, Excalibur Publications, 2002.

[36] "Webley top-break revolvers" (http://world.guns.ru/handguns/hg91-e.htm). world.guns.ru. .

[37] Bishop, Chris (1998). *Guns in Combat*. Chartwell Books, Inc. ISBN 0-7858-0844-2.

[38] http://americanrifleman.org/ArticlePage.aspx?id=1359&cid=28

Smith & Wesson Victory Model

Smith & Wesson Military & Police	
Lend-Lease M&P dating from World War II, missing lanyard ring	
Type	Service revolver
Place of origin	United States
Service history	
Used by	See Users
Wars	World war I, World War II, Korean War, Vietnam War
Production history	
Designed	1899
Manufacturer	Smith & Wesson
Variants	.38 Hand Ejector, M&P Model of 1905, Victory Model, Model 10
Specifications	
Weight	~ 34 oz. (907 g) with standard 4" (102 mm) barrel (unloaded)
Length	254 mm, 260 mm or 286 mm
Caliber	.38 Special .38/200
Action	Double action
Muzzle velocity	305 m/s (755 Feet Per Second) (*.38 spl*) 198 m/s (685 Feet Per Second) (*.38/200*)
Feed system	6-round cylinder
Sights	Blade front sight, notched rear sight

The **Smith & Wesson Model 10**, previously known as the **Smith & Wesson Military & Police**, and for those produced during World War II, the **Smith & Wesson Victory Model**, is a .38-caliber, six-shot handgun initially developed in 1899 as the Smith & Wesson **.38 Hand Ejector** model. This model in all its incarnations has been in production since 1899.

"The .38in *Military and Police* Model 10 has historically been the mainstay of the Smith & Wesson Company, with some 6,000,000 of this general type produced to date. It has been described as the most successful handgun of all time, and the most popular centerfire revolver of the 20th Century."[1]

History

The original Military & Police Model of 1899 was built around the .38 S&W Special round--a slightly elongated improvement on the .38 Long Colt with increased bullet weight (158 grains) and increase in powder charge from eighteen to twenty-one grains of black powder. The round's full name is actually .38 S&W Special. A number of the first models were chambered for .38 Long Colt to satisfy a government order.[2] Serial numbers ranged from number 1 in the series to 20,975 at which point (1902), the model underwent substantial changes.[3] , in particular the S&W Military & Police Model of 1905 (mfg 1905 - 1942) chambered in .38 Special.

Changes include major modification and simplification of the internal lockwork and addition of a barrel- mounted locking lug to engage the here-to-fore free standing ejector rod. The 4th change of April 20, 1915 had enlarged service sights that quickly became a standard across the service revolver segment of the industry. Heat treating of cylinders began in 1919.[4]

The first model M&P of 1899. The ejector rod is free-standing lacking the front barrel latch of later models

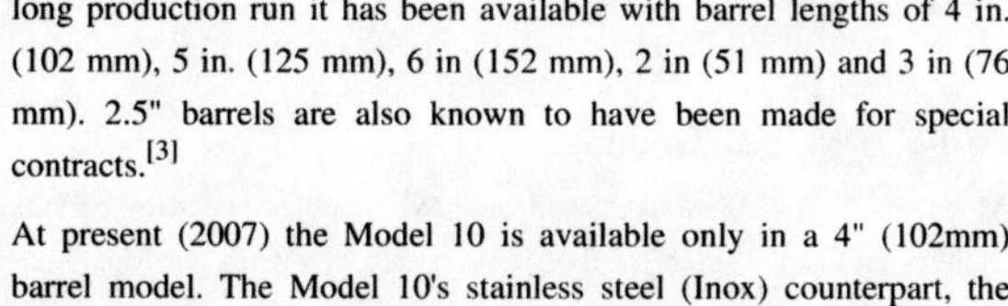

The Model 10 is a fixed-sight revolver with a fluted cylinder. Over its long production run it has been available with barrel lengths of 4 in. (102 mm), 5 in. (125 mm), 6 in (152 mm), 2 in (51 mm) and 3 in (76 mm). 2.5" barrels are also known to have been made for special contracts.[3]

At present (2007) the Model 10 is available only in a 4" (102mm) barrel model. The Model 10's stainless steel (Inox) counterpart, the Smith & Wesson Model 64, is available in either a 4" (102 mm) or a 3" (76 mm) barrel.

Victory model

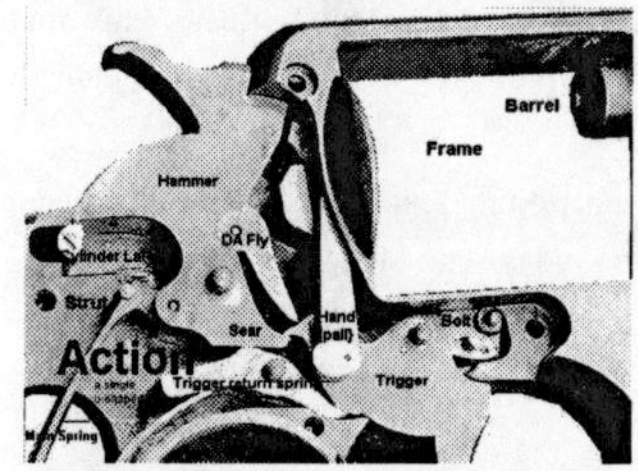

The lockwork of the first model differed substantially from subsequent versions. Note that the trigger return spring is a flat leaf rather than a coil spring powered slide, unlike the variations dating from 1905 onwards.

The S&W Model 10 military revolvers produced from 1940 to 1945 had serial numbers with a "V" prefix, and were known as the **Smith & Wesson Victory Model**. It is noteworthy that early Victory Models did not always have the V prefix. During World War II, huge numbers - over 570,000 - of these pistols, chambered in the British .38/200 caliber already in use in the Enfield No 2 Mk I Revolver and the Webley Mk IV Revolver, were supplied to the United Kingdom, Canada, Australia, New Zealand, and South Africa under the Lend-Lease program. Most Victory Models sent to Britain were fitted with 4" (102 mm) or 5" (127 mm) barrels, though a few early versions had 6" (150 mm) barrels.[5] [6] In general, most British and Commonwealth forces expressed a preference for the .38/200 Smith & Wesson over their standard Enfield revolver.[7]

The Victory Model was also used by United States forces during WWII, being chambered in the well-known and popular .38 Special cartridge. The Victory Model was a standard-issue sidearm for US Navy and Marine aircrews, and was also used by guards at factories and defense installations throughout the United States during the war. Some of these revolvers remained in service well into the 1990s with units of the US Armed Forces, including the Coast Guard. Some Lend-Lease Victory Model revolvers originally chambered for the

British .38/200 were returned to the U.S. and rechambered to fire the more popular and more powerful .38 Special ammunition, and such revolvers are usually so marked on their barrels. Rechambering of .38-200 cylinders to .38 Special results in oversized chambers which may cause problems.

The finish on Victory Models is typically a sandblasted and parkerized finish, which is noticeably different from the higher-quality blue or nickel/chrome finishes usually found on commercial M&P/Model 10 revolvers. Other distinguishing features of the Victory Model revolver are the lanyard loop at the bottom of the grip frame, and the use of smooth (rather than checkered) walnut grip panels. However some early models did use a checkered grip. Most notably the pre-1942 manufacture.[8]

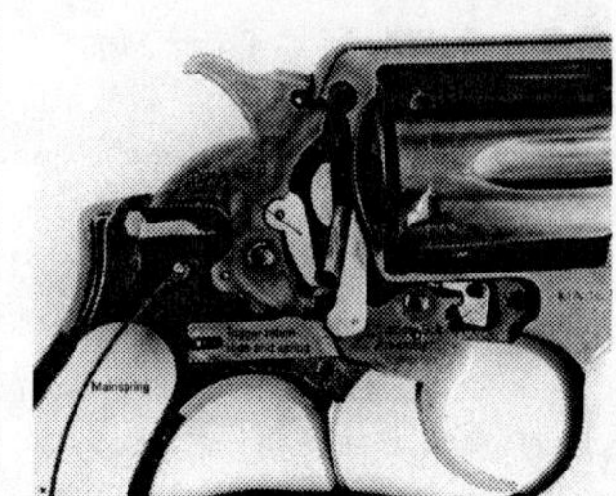

The M&P 1905 fourth change of 1915. The lock mechanism remained principally unchanged after this model.

Post-World War II models

After World War II ended, Smith and Wesson returned to manufacturing the M&P series. Along with cosmetic changes and replacement of the frame fitting grip with the Magna stocks, the spring-loaded hammer block safety gave way to a cam-actuated hammer block that rode in a channel in the side plate (Smith 1968). In 1957, Smith and Wesson began a convention of using numeric designations to distinguish their various models of handguns. The M&P was renamed the Model 10.

The M&P/Model 10 has been available in both blued steel finish and nickel finish for most of its production run. The model has also been offered throughout the years with both the round butt and square butt (i.e. grip patterns). Beginning with the Model 10-5 series in the late 1960s, the tapered barrel and its trademark 'half moon' front sight (as shown in the illustrations on this page) were replaced by a straight bull barrel and a sloped milled ramp front sight. Late model Model 10s are capable of handling any .38 Special cartridge produced today up to and including +P+ rounds.

As its name suggests, the S&W Military & Police revolver was developed for military and police use. In this regard it has been enormously successful, with the Model 10 remaining in production today. The Model 10 has also enjoyed popularity with civilian shooters in countries that allow private ownership of handguns.

.357 Magnum variations

After a small prototype run of Model 10-6 revolvers in .357 Magnum caliber, Smith and Wesson introduced the Model 13 heavy barrel in carbon steel and then the Model 65 in stainless steel. Both revolvers featured varying barrel weights and lengths—generally three and four inches with and without underlugs (shrouds). Production dates begin in 1974 for the Model 13 and end upon discontinuation in 1999. The Model 65 was in production from 1972-1999.[3] Both the blued and stainless models were popular with police and FBI and a variation of the Model 65 was marketed in the Lady Smith line from 1992-1999. Circa 2005, S&W discontinued all K-frame .357 Magnums in favor of the L frame size.

Semi-automatic M&P

As of 2005, Smith & Wesson produces a polymer-framed semi-automatic handgun in 9 mm, .40 S&W, and .357 SIG under the M&P name. In 2007, a .45 ACP version of the semi-automatic M&P was introduced.

Users

- Australia Currently used by Victoria Police (Being replaced by Smith & Wesson M&P)
- Austria (Austrian Police during occupation after WW2)
- Canada
- Hong Kong Used by Royal Hong Kong Police Constables
- Ireland (Garda Siochana)
- Israel
- Macau
- Malaysia
- New Zealand
- Norway Used by the Home Guard until 1986. Used by the **Norwegian Police Service** (Norwegian: *Politi- og lensmannsetaten*) until 2008.
- Pakistan
- Peru (Peruvian National Police)
- South Africa
- United Kingdom
- United States Used by United States Army and United States Marine Corps

See also

- Colt Commando
- Enfield revolver
- M1917 revolver
- Service pistol
- Smith & Wesson

External links

- World Guns page [9]

References

[1] The History of Smith & Wesson Firearms (http://books.google.com/books?id=5jnqqkW85u8C) Dean K. Boorman, 2002, pp 46
[2] Cumpston, Mike (2003-01-16). "The First M&P" (http://www.gunblast.com/Cumpston_SW-MP.htm). Gunblast.com. . Retrieved 2008-05-02.
[3] Supica, Jim; and Richard Nahas (2001). *Standard Catalog of Smith & Wesson*. Iola WI: Krause Publications. p. 1068.
[4] Smith, W.H.B (1968). *Book of Pistols and Revolvers* (7th Edition ed.). Harrisburg: Stackpole Books.
[5] Shore, C. (Capt), *With British Snipers to the Reich*, Paladin Press (1988), p. 55
[6] Dunlap, Roy, *Ordnance Went Up Front*, Samworth Press (1948), p. 142
[7] Shore, C. (Capt), *With British Snipers to the Reich*, Paladin Press (1988), p. 202
[8] Hunter, Hunter (2009), "S&W Victory & Colt Commando Revolvers", *American Rifleman* (Fairfax, Virginia: National Rifle Association of America) (Vol. 157, No. 6, June 2009): 36–37, ISSN 0003-083X
[9] http://world.guns.ru/handguns/hg76-e.htm

Colt New Service

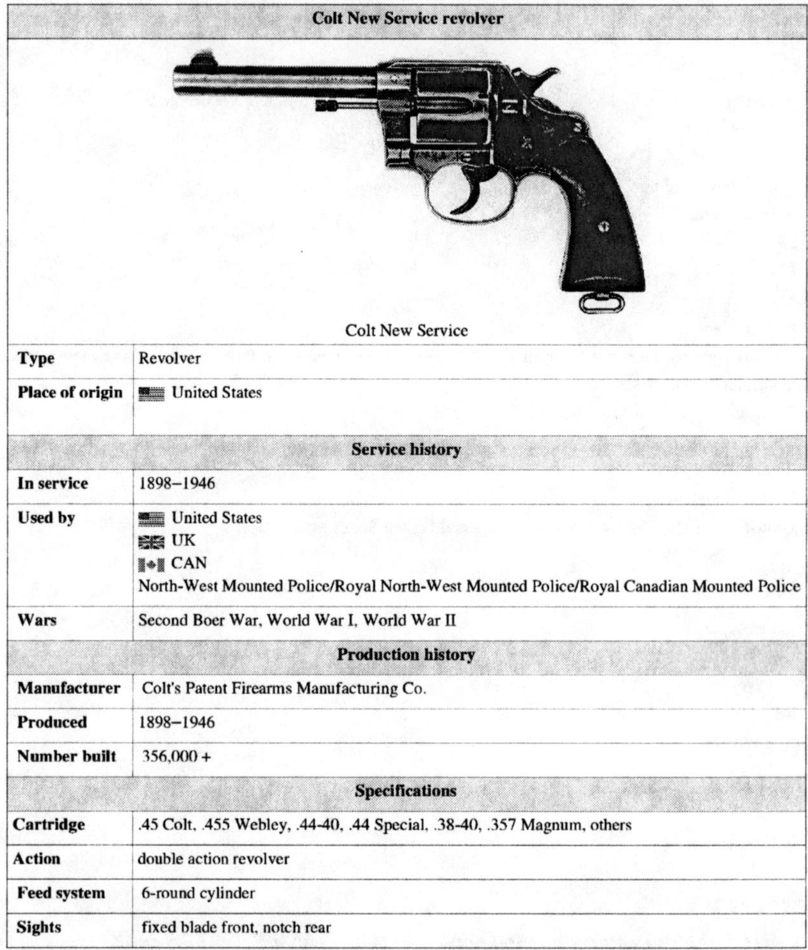

Colt New Service revolver	
Colt New Service	
Type	Revolver
Place of origin	United States
Service history	
In service	1898–1946
Used by	United States UK CAN North-West Mounted Police/Royal North-West Mounted Police/Royal Canadian Mounted Police
Wars	Second Boer War, World War I, World War II
Production history	
Manufacturer	Colt's Patent Firearms Manufacturing Co.
Produced	1898–1946
Number built	356,000 +
Specifications	
Cartridge	.45 Colt, .455 Webley, .44-40, .44 Special, .38-40, .357 Magnum, others
Action	double action revolver
Feed system	6-round cylinder
Sights	fixed blade front, notch rear

The **Colt New Service** was a double-action revolver made by Colt from 1898 until c.1946. It was adopted by the U.S. Armed Forces in .45 Colt as the Model 1909 U.S. Army, Marine Corps Model 1909, Model 1909 U.S. Navy and in .45 ACP as the Model 1917 U.S. Army.[1] The Model 1917 was created to supplement inadequate stocks of M1911 pistols during World War One.[2]

In 1899 Canada acquired a number of New Service revolvers (chambered in .45 Colt) for Boer War service, to supplement its existing Model 1878 Colt Double Action revolvers in the same caliber.[3] In 1904/5 the North-West Mounted Police in Canada also adopted the Colt New Service to replace the less-than satisfactory Enfield Mk II revolver in service since 1882.[4]

New Service revolvers, designated as **Pistol, Colt, .455-inch 5.5-inch barrel Mk. I**, chambered for the .455 Webley cartridge were acquired for issue as "substitute standard" by the British War Department during World War One.[5]

British Empire Colt New Service Revolvers were stamped *"NEW SERVICE .455 ELEY"* on the barrel[6] , to differentiate them from the .45 Colt versions used by the US (and Canada).

The Colt New Service was a popular revolver with British Officers, and many of them had privately purchased their own Colt New Service revolvers in the years prior to World War I as an alternative to the standard-issue Webley Revolver. 60,000 Colt New Service revolvers were supplied to British Empire and Canadian forces during WW I, and they continued to see service until the end of World War II.[6]

See also

- Antique Guns

References

- Chamberlain & Taylerson, W.H.J. & A.W.F. (1989). *Revolvers of the British Services, 1854-1954*. Bloomfield, ON (Canada) and Alexandria Bay, NY (USA). ISBN 0-919316-92-1.
- Law, Clive M. (1994). *Canadian Military Handguns, 1855-1985*. Bloomfield, ON (Canada) and Alexandria Bay, NY (USA). ISBN 0-88855-008-1.
- Maze, Robert J. (2002). *Howdah to High Power: A Century of Breechloading Service Pistols (1867-1967)*. Tucson, AZ (USA). ISBN 1-880677-17-2.
- Murphy, Bob (1985). *Colt New Service Revolvers*. Aledo, Illinois (USA).
- Phillips & Klancher, Roger F. & Donald J. (1982). *Arms & Accoutrements of the Mounted Police, 1873-1973*. Bloomfield, ON (Canada) and Alexandria Bay, NY (USA). ISBN 0-919316-84-0.

External references

- Guns and Ammo Magazine article on Colt New Service Revolver [7]

References

[1] Murphy, Bob. (1985). *Colt New Service Revolvers*. World-Wide Gun Report Inc. pg.25-30

[2] Ibid: pg.31

[3] Law (1994) pp.28-30

[4] Phillips & Klancher (1982) pp.21ff

[5] Chamberlain & Taylerson (1989) p. 54ff; Maze (2002) p.85

[6] Maze (2002) p.84

[7] http://gunsandammomag.com/cs/Satellite/IMO_GA/Guide_C/Colt+New+Service+Revolver

Smith & Wesson No. 3 Revolver

Smith & Wesson No. 3 Revolver	
Smith & Wesson No. 3 Third Model Russian	
Type	Service Revolver
Place of origin	United States
Service history	
Used by	U.S. Military, Imperial Russia, Argentina
Wars	Indian Wars Russo-Turkish War (1877–1878) Spanish American war Philippine–American War
Production history	
Manufacturer	Smith & Wesson
Specifications	
Caliber	.44 Russian, .44 S&W American
Action	Single Action
Feed system	6-round cylinder
Sights	fixed front post and rear notch

The **Smith & Wesson model 3** (AKA the **Smith & Wesson Russian Model** and the **Schofield Revolver**) was a single-action,cartridge-firing, top-break revolver produced by Smith & Wesson from 1870 to the late 1880s, and again recently as a reproduction by Smith & Wesson themselves, Armi San Marco, and Uberti.

It was produced in several variations and sub-variations, including both the aforementioned "Russian Model" (so named because it was supplied to military of Tsarist Russia)[1] and the "Schofield" model, named after Major George W. Schofield [2] who made his own modifications to the Model 3 to meet his perceptions of the Cavalry's needs, which Smith & Wesson incorporated into an 1875 design they named after the Major, planning to obtain significant military contracts for the new revolver.

The S&W model 3 was originally chambered for the .44 S&W American and .44 Russian cartridges, and typically did not have the cartridge information stamped on the gun (as is standard practice for most commercial firearms). model 3 revolvers were also later produced in an assortment of calibres including .44 Henry Rimfire, .44-40, .32-44, .38-44, and .45 Schofield.[3]

Russian Model

Smith & Wesson produced large numbers of the Model 3, in three distinct models, for Tsarist Russia by special order. The first was the 1st Model Russian (the original order design), and the Russian Ordnance Inspector mandated a number of improvements to the design, resulting in the 2nd Model Russian, with a final revision to the Russian design being known as the 3rd Model Russian.

Smith & Wesson nearly went bankrupt as a result of their Russian Contract production, as the Tsarist government assigned a number of engineers and gunsmiths to reverse-engineer the Smith & Wesson design, and then began to produce copies of the revolver—both in their own arsenal at Tula and by contracting other manufacturers in Germany and elsewhere in Europe to manufacture copies of the revolver (a common practice at the time—Webley & Scott's British Bulldog revolver was widely copied by European and American gunsmiths).

The Russian and European copies of the S&W model 3 revolver were generally of very high quality, but considerably cheaper than the S&W produced revolvers. This led to the Tsarist government cancelling the order for significant quantities of Smith & Wesson–made revolvers (which Smith & Wesson had already produced), and delaying (or refusing) payment for the handguns that had already been delivered.

Schofield Revolver

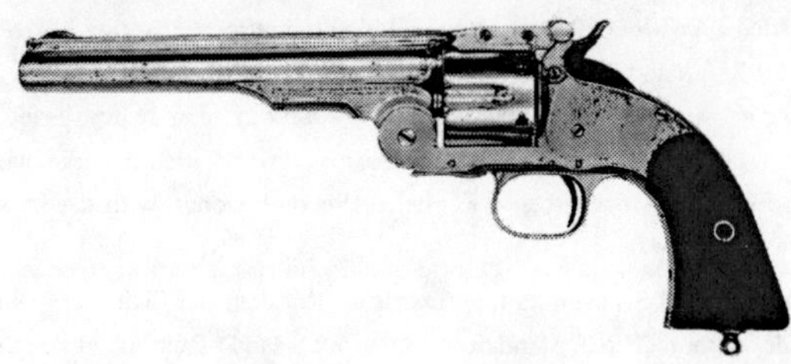

Smith & Wesson Model No. 3 Schofield Revolver

The US Army adopted the .44 S&W American calibre Smith & Wesson model 3 revolver 1870, making the model 3 revolver the first standard-issue cartridge-firing revolvers in US service. Most military pistols up until that point were black powder cap and ball revolvers, which were (by comparison) slow, complicated, and susceptible to the effects of wet weather.

In 1875 the US Ordnance Board granted Smith & Wesson a contract to outfit the military with model 3 revolver incorporating the design improvements of Major George W. Schofield (known as the "Schofield revolver"), providing they could make the revolvers work with the .45 Colt (AKA ".45 Long Colt") ammunition already in use by the US military. Smith & Wesson instead developed their own, slightly shorter .45 caliber round, the .45 Schofield, otherwise known as the .45 S&W. When it became obvious in the field that the two cartridges would not work interchangeably in the Schofield (although they both worked in the Colt), the U.S. Government adopted the shorter .45 Schofield cartridge as the standard cartridge. Despite the change, old stocks of the longer .45 Colt rounds in the supply line caused the Army to drop most of the Schofields and continue with the Colt. Major Schofield had patented his locking system and earned a payment on each gun that Smith and Wesson sold, and at the time his older brother, John M. Schofield, was the head of the Army Ordnance Board and the political situation may have been the main issue for the early end of army sales.

Many of the S&W model 3 Schofield revolvers saw service in the Indian Wars, and there are reports of them in use as late as the Spanish-American War and Philippine-American War. Like the other Smith and Wesson Model 3's, they were also reportedly popular with lawmen and outlaws in the American West, and were reportedly used by Jesse James, John Wesley Hardin, Pat Garrett, Theodore Roosevelt, Virgil Earp, Billy the Kid, and many others.

While the standard barrel length was 7", many Schofields were purchased as surplus by distributors, and had the barrels shortened to 5", and were refinished in nickel.

After the Spanish American War in 1898, the US Army sold off all their surplus Schofield revolvers. The surplus Schofield revolvers were reconditioned by wholesalers and gunsmiths (at professional factory-quality level) with a considerable number offered for sale on the commercial market with a 5 inch barrel as well as the standard size barrel of 7 inches.

Of the most notable purchasers of these reconditioned model 3 Schofield revolvers was Wells Fargo and Company, who purchased the revolvers for use by Wells Fargo Road Agents and had the barrels shortened to a more concealable 5 inch length. These revolvers were then inspected by Wells Fargo armorer and uniquely stamped "W.F. & Co" or "Wells Fargo & Co", along with the original Smith & Wesson serial number re-stamped alongside the Wells Fargo stamping on the flat part of the barrel just forward of the barrel pivot as well as re-stamping any part of each revolver which had not originally been stamped or stamped in a location that would be difficult to view the serial number, when needed.

The Wells Fargo Schofield revolvers became so popular with collectors from the 1970s onwards that the unique Wells Fargo markings were being "counterfeited" or "faked" by unscrupulous sellers to enhance the value of other similar versions that had not been genuinely owned by Wells Fargo & Co. There are more "fake" Wells Fargo marked Schofield revolvers than genuine ones in existence and, accordingly, a collector interested in purchasing a "Wells Fargo" Schofield revolver would be well advised to have a pre-purchase inspection and verification performed by an expert who specializes in this model.

The Schofield was produced in two versions, the First Model Schofield and the Second Model Schofield. The First Model Schofield has a latch configuration that is rather pointed at the top and has a circle around the screw head at the bottom, whereas the Second Model latch has a large raised circle at the top of the latch. One of General Schofield's revisions and improvements to the predecessor Model 3 Revolvers included mounting the spring loaded barrel catch on the frame as opposed to the standard Smith and Wesson Model 3 has the latch mounted on the barrel rather than the frame. In the previous engineering the posts of the frame would wear out after heavy usage. Schofield's improvement called for heat treated, replaceable components at this sensitive "wear" area of the catch and latch. Serial number range also will give an indication of whether it is First or Second Model, with the s/n's changing from the First Model to the Second Model at a little over 3,000.

In 1877, S&W discontinued production of its other Model 3's such as the American, Russian, and Schofield—in favor a new improved design called the New Model number Three. Standard chambering was .44 Russian, although other calibers were offered on special order or in related models such as the .44-40 Frontier Model, the .32-44 & .38-44 Target Models, and the very rare .38-40 Winchester Model.

Modern reproductions

Modern reproductions of the Smith & Wesson model 3 Revolver are made by a number of companies, including (most notably) Smith & Wesson themselves, as well as the Italian arms-makers Uberti and Armi San Marco.

Smith and Wesson

Smith & Wesson manufacture a modern reproduction of the original model 3 Revolvers. Re-introduced at the 2000 SHOT Show, and touted as being a "true" reproduction, side-by-side comparison of an original with the pre-production gun showed that the new version is slightly beefier than the original around the barrel and topstrap, though not as much as on the Navy Arms guns. Changes in the internal lock mechanism were also made. It appears from the photos that the firing pin in the S&W Model is frame-mounted instead of being an integral part of the hammer, a modern safety feature - with a "transfer bar" as a practical safety catch in a revolver - preventing accidental discharge if dropped.

Uberti/Armi San Marco

The Uberti version, imported by Navy Arms, had external dimensions generally similar to the original 2nd Model Schofield, but the barrel and topstrap are considerably thicker, for additional strength. As with the Armi San Marco model, the Navy Arms/Uberti model 3 revolver has a lengthened cylinder to accommodate .45 Long Colt and .44-40 cartridges. Although there were some problems with the locking latch angles in early guns, these were generally corrected or the guns replaced. European reproduction model 3 revolvers have changes made to their lockwork to meet import regulations.[4]

References

[1] Boorman, Dean K. (2002). *The History of Smith & Wesson Firearms* (http://books.google.com/books?id=5jnqqkW85u8C&pg=PA32&lpg=PA32#v=onepage&q&f=false). Guilford, Conn.: Globe Pequot Press. p. 32. ISBN 1585747211. . Retrieved April 23, 2010.

[2] Boorman. *The History of Smith & Wesson Firearms* (http://books.google.com/books?id=5jnqqkW85u8C&pg=PA33&lpg=PA33#v=onepage&q&f=false). p. 33. . Retrieved April 23, 2010.

[3] Boorman. *The History of Smith & Wesson Firearms* (http://books.google.com/books?id=5jnqqkW85u8C&pg=PA31&lpg=PA31#v=onepage&q&f=false). p. 31. . Retrieved April 23, 2010.

[4] "Uberti Top Break: No. 3 New Model Russian, No. 3 2nd Model" (http://www.uberti.com/firearms/top_break.php). A. Uberti. . Retrieved April 23, 2010.

Double action only

A **trigger** is a mechanism that actuates the firing sequence of firearms, or a power tool. Triggers almost universally consist of levers or buttons actuated by the index finger. Rare variations use the thumb or weak fingers to actuate the trigger. Examples are the M2 Browning machine gun and the Springfield Armory M6 Scout.

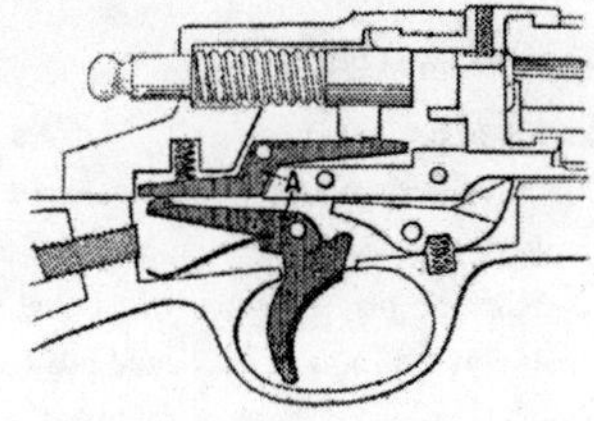

Trigger mechanism in a bolt action rifle.

Function

Firearms use triggers to initiate the firing of a cartridge in the firing chamber of the weapon. This is accomplished by actuating a striking device through a combination of spring and kinetic energy operating through a firing pin to strike and ignite the primer. There are two primary types of striking mechanisms, hammers and strikers. Hammers are spring-tensioned masses of metal that pivot on a pin when released and strike a firing pin to discharge a cartridge. Strikers are, essentially, spring-loaded firing pins that travel on an axis in-line with the cartridge eliminating the need for a separate hammer. The connection between the trigger and the hammer is generally referred to as the sear surface. Variable mechanisms will have this surface directly on the trigger and hammer or have separate sears or other connecting parts.

Mechanisms

There are numerous types of trigger mechanisms. They are categorized according to which functions the trigger is to perform. In addition to releasing the hammer or the striker, a trigger may cock the hammer or striker, rotate a revolver's cylinder, deactivate passive safeties, select between semi-automatic and full-automatic fire such as the Steyr AUG, or pre-set a 'set trigger'. Most modern firearms use the trigger to deactivate passive safeties but this does not change how they are identified.

Single-action

A **single-action** (SA) trigger performs the *single action* of releasing the hammer or striker to discharge the firearm each time the trigger is pulled.[1] Almost all rifles and shotguns use this type of trigger.[1] Single-action semi-automatic pistols require that the hammer be cocked before the first round can be fired, although most designs cock the hammer as part of the loading process (e.g., the act of inserting the magazine and operating the slide mechanism to chamber the first round also cocks the hammer or striker into the ready-to-fire position).[2] Once the first round is fired the automatic movement (recoil) of the slide cocks the hammer for each subsequent shot. The pistol, once cocked, can be fired by pulling the trigger once for each shot until the magazine is empty. The M1911, Browning Hi-Power, Springfield XD, and Smith & Wesson M&P are single-action pistols that function in this manner.[2] Single action revolvers require the hammer to be cocked by hand every time the weapon is fired.

Double-action/single-action

A **double-action/single-action** (DA/SA) firearm combines the features of both mechanisms. Often called **traditional double action,** these terms apply almost exclusively to semi-automatic handguns. The function of this trigger mechanism is identical to a DA revolver. However, the firing mechanism automatically cocks the hammer or striker after the gun is fired. This mechanism will cock and release the hammer when the hammer is in the down position but on each subsequent shot, the trigger will function as a single action. The Mateba Autorevolver is a hybrid revolver that functions on a DA/SA system. However, it is different in function than either a conventional revolver or semi-automatic pistol. The Beretta 92 is a good example of a DA/SA semi-automatic pistol. On many DA/SA pistols (including the Beretta) there is the option to cock the hammer before the first shot is fired. This removes the heavy pull of the double-action. Also, there is often a de-cocker to return the pistol to double-action.

Double-action

A **double-action,** also known as double action only (DAO) to prevent confusion with DA/SA designs, is similar to a DA revolver trigger mechanism however there is no single action function. A Sigarms DAK trigger is a good example of this and a sure way to tell if a trigger meets this design is to see if there is second-strike capability. For semi-automatic pistols with a traditional hammer (that employ only the double action function of the trigger), the hammer will return to its decocked position after each shot. Subsequent shots require the double action trigger firing sequence. For striker-fired pistols such as the Taurus 24/7, the striker will remain in the rest position through the entire reloading cycle. This term applies mostly to semi-automatic handguns; however, the term can also apply to some revolvers such as the Smith & Wesson Centennial, the Type 26 Revolver, and the Enfield No. 2 Mk I revolvers, in which there is no external hammer spur. Glock and Kahr semi-automatic pistols are not DA (or DAO) pistols because the striker is "cocked" to an intermediate position by the operation of the slide and they cannot be re-activated by pulling the trigger a second time.

Release trigger

A release trigger releases the hammer or striker when the trigger is released by the shooter, rather than when it is pulled.[3] Release triggers are largely used on shotguns intended for trap shooting.

Set Trigger

A set trigger allows a shooter to have a greatly reduced trigger pull (the resistance of the trigger) while maintaining a degree of safety in the field. There are two types: Single Set and Double Set.

Single set trigger

A Single Set Trigger is usually one trigger that may be fired with a conventional amount of trigger pull weight or may be 'set' by usually pushing forward on the trigger. This takes up the creep in the trigger and allows the shooter to enjoy a much lighter trigger pull.

Double set trigger

As above, a double set trigger accomplishes the same thing, but uses two triggers: one sets the trigger and the other fires the weapon. Set triggers are most likely to be seen on customized weapons and competition rifles where a light trigger pull is beneficial to accuracy.

Double set triggers can be further classified by phase.[4] A double set, single phase trigger can only be operated by first pulling the set trigger, and then pulling the firing trigger. A double set, double phase trigger can be operated as a standard trigger if the set trigger is not pulled, or as a set trigger by first pulling the set trigger. Double set, double phase triggers offer the versatility of both a standard trigger and a set trigger.

Pre-set (striker or hammer)

Pre-set strikers and hammers apply only to semi-automatic handguns. Upon firing a cartridge or loading the chamber, the hammer or striker will rest in a partially cocked position. The trigger serves the function of completing the cocking cycle and then releasing the striker or hammer. While technically two actions, it differs from a double-action trigger in that the trigger is not capable of fully cocking the striker or hammer. It differs from single action in that if the striker or hammer were to release, it would generally not be capable of igniting the primer.

Examples of pre-set strikers are the Glock, Kahr Arms, and Ruger SR9 pistols.

Examples of pre-set hammers are the Kel-Tec P-32 and Ruger LCP pistols.

Pre-set hybrid

Pre-set hybrid triggers are similar to a DA/SA trigger in reverse. The first pull of the trigger is pre-set. If the striker or hammer fail to discharge the cartridge, the trigger may be pulled again and will operate as a Double Action Only (DAO) until a malfunction is cleared or the cartridge discharges. This allows the operator to attempt to fire a cartridge after a misfire malfunction. The Taurus PT 24/7 Pro pistol (not to be confused with the first-generation 24/7 which was a traditional pre-set) offers this feature as of 2006.

Relative merits

Each trigger mechanism has its own merits. Historically, the first type of trigger was the single action.[2] This is the simplest mechanism and generally the shortest, lightest, and smoothest pull available.[2] The pull is also consistent from shot to shot so no adjustments in technique are needed for proper accuracy. On a single-action revolver, for which the hammer must be manually cocked prior to firing, an added level of safety is present. On a semi-automatic, the hammer will be cocked and made ready to fire by the process of chambering a round, and as a result an external safety is sometimes employed.

Double action triggers provide the ability to fire the gun no matter whether the hammer is cocked or uncocked. This feature is desirable for military, police, or self-defense pistols. The primary disadvantage of any double-action trigger is the extra length the trigger must be pulled and the extra weight required to overcome the spring tension of the hammer or striker.

DAO firearms attempt to solve the problems with DA/SA mechanisms by making every shot a double-action shot. Because there is no difference in pull weights, training and practice are simplified. Additionally, negligent discharges are reduced because of the heavier trigger pulls. This is a particular advantage for a police pistol. These weapons also generally lack any type of external safety. DAO is common among police agencies and for small, personal protection firearms. The primary deficiency is that accurate fire is difficult due to the additional trigger weight and travel required for each shot.

DA/SA pistols are versatile mechanisms. These complex firearms generally have a manual safety that may or may not serve to decock the hammer. Some have a decocking lever and a manual safety as well. As a disadvantage, these levers are often intermingled with other levers such as slide releases and take-down levers with variables that become confusing. Training will largely overcome this weakness. One other disadvantage is the difference between the first double-action pull and subsequent single-action pulls (if the hammer is not set before the first shot by the user).

Pre-set triggers, only recently coming into vogue, offer what some would consider an optimum balance of pull weight, trigger travel, safety, and consistency. Glock pioneered this trigger and many other manufacturers have followed suit. The primary disadvantage of the pre-set trigger is that pulling the trigger a second time after a failure to fire will not re-strike the primer. In normal handling of the firearm, this is not an issue; loading the gun requires that the slide be retracted, pre-setting the striker. Clearing a malfunction also usually involves retracting the slide.

References

[1] "Trigger Analysis" (http://www.ballistics-experts.com/Forensic ballistics/Trigger analysis.htm). *Ballistics-Experts.com.* .
[2] "Trigger Options of the Semi-Automatic Service Pistol" (http://www.chuckhawks.com/trigger_options.htm). *Chuckhawks.com.* .
[3] U.S. Patent 2027950 (http://www.google.com/patents?vid=2027950)
[4] "Trigger Function and Terminology" (http://members.aye.net/~bspen/triggerterm.html). *Spencer, B. E..* .

Colt Official Police

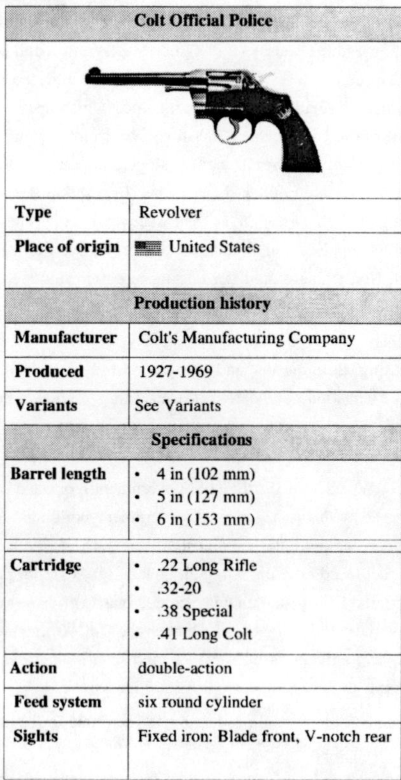

Colt Official Police	
Type	Revolver
Place of origin	United States
Production history	
Manufacturer	Colt's Manufacturing Company
Produced	1927-1969
Variants	See Variants
Specifications	
Barrel length	• 4 in (102 mm) • 5 in (127 mm) • 6 in (153 mm)
Cartridge	• .22 Long Rifle • .32-20 • .38 Special • .41 Long Colt
Action	double-action
Feed system	six round cylinder
Sights	Fixed iron: Blade front, V-notch rear

Introduced to the firearms market in 1927, The **Colt Official Police** is a medium frame, six-shot, double-action revolver with a six round cylinder, primarily chambered for the .38 Special cartridge, and manufactured by the Colt's Manufacturing Company. The intended specific target market for the Official Police was mainly law enforcement agencies and it became one of the best selling police firearms of all time, eventually in the 1950s coming to exemplify typical peace officer weaponry.[1] The Official Police was also used by various U.S. and allied military forces during World War II.

Development and history

As the 20th Century began, the older .32 caliber revolvers which had been standard-issue for the majority of American police departments began to be phased out in favor of the larger-bore .38 caliber. In 1908 Colt introduced a sleek and modernized revolver they dubbed the Army Special, or "New Army" which in the powerful (for the time) and popular .38 Special quickly became the issue service revolver of many departments.[1] [2] [3] During the same period, revolvers began to fall out of favor with the U.S. Military, especially after the adoption on the U.S. Model of 1911 semi-automatic pistol.[1] As military sales of their revolvers dropped off, Colt searched for an alternative market and realized the popularity and strong sales of their product line with civilian law enforcement agencies could form their replacement market.[2]

By 1927 the overwhelming sales of two popular models, the New Army and Colt Police Positive, had assured Colt's dominance of the law enforcement firearms market.[1] [2] Colt's marketing strategy was further fine-tuned by making a few superficial alterations to the New Army revolver and then renaming it as the "Official Police" model.[1] [2] The changes included adding checkering to the trigger and cylinder latch, matting the topstrap of the frame and widening the rear sight groove. Colt also upgraded the quality of the gun's finish from a dull blued finish to a highly polished blued surface.[1] [3] In 1930, Colt scored a marketing coup when they publicized that their Official Police model could easily handle the firing of heavily-loaded .38 rounds intended for competitor Smith & Wesson's new large N-frame revolver, the .38-44, none of the comparable S&W revolvers could manage this feat.[1] [2] By 1933 the Colt sales catalog listed many law enforcement agencies as having adopted the OP as a sidearm, including the New York City, Los Angeles, Chicago, and Kansas City police departments. In addition many state police organizations and even the Federal Bureau of Investigation chose the OP as their issue revolver.[1] The U.S. Army also bought some of the revolvers, issuing them to military police and to federal agencies in need of a revolver for their armed agents, such as the Treasury Department, Coast Guard, and the Postal Inspection Service. Many Official Police revolvers were also bought by the police forces and militaries of various South American countries.[1] [3]

Between May 1940 and June 1941, 49,764 Official Police revolvers in .38 New Police or .38/200 caliber were purchased by the British Purchasing Commission and shipped to the United Kingdom for use by British and Commonwealth armed forces as a substitute standard sidearm. These revolvers bore British military acceptance markings and had a 5" barrel; the butt was fitted with a military-style lanyard ring. Most of these OP revolvers were assembled from commercial-grade parts made before 1942.[4]

When the U.S. became involved in World War II , the U.S. government requested contracts to supply .38 revolvers required for arming officers charged with the security of government buildings against sabotage or theft. When government purchasing officials objected to the unit cost of the Official Police revolver, Colt responded by simplifying the OP. Savings were achieved by eliminating all unnecessary exterior polishing operations, substituting a smooth-face trigger and hammer, and fitting the gun with molded plastic grips instead of wood. Rather than bluing, the revolver was given a dull parkerized finish treatment. Dubbed the Colt "Commando", the weapon was purchased in .38 Special caliber and used to arm units of Military police, and security agencies, as well as limited clandestine issue to agencies involved in overseas espionage and military intelligence.[1] [2] [5]

After the Allied victory, Colt resumed commercial production and returned to the prewar polished blued finish, but retained the plastic grips which they labeled "Coltwood" until 1954 when the checkered wooden grips were reintroduced.[1] During the postwar period, Colt fell on difficult financial times and the company introduced few new models. At Smith & Wesson, both output and new model civilian and police sales improved, and the sales margin gap between the two corporations progressively tightened. Finally in the 1960s S&W took over the lead.[1] [2] A contributing factor to this change may have been Smith & Wesson's generally lower cost per unit, accompanied by a double-action trigger pull on their military & police model that was preferred by many agencies teaching the new combat-oriented double-action revolver training.[1] [2] Colt announced the discontinuation of the Official Police in 1969, stating that competitive production of the design was no longer economically feasible.[1] [2] With a total production of over 400,000 pistols, the Official Police ranks as one of the most successful and well-liked handguns ever made.[1]

Features

The Official Police was machined of fine carbon steel, with bright royal Colt blued as well as nickel plated finishes, and was offered in 4", 5", and 6" barrels. Built on Colt's .41 or "I" frame, it was manufactured in a variety of chamberings including .22 LR, .32-20 (discontinued in 1942), .41 Long Colt (discontinued in 1938), and the most common and popular, the .38 Special.[1] [2] Colt's "Positive Lock" firing pin block safety was a standard feature of the revolver, preventing the firing pin from striking the primer unless the trigger was deliberately pulled.[1] The pistol's sights consisted of a blade front with a fixed iron open rear sight, which was a simple V-notch shaped groove milled into the revolver's top strap. The top strap had a matte finish to reduce glare down the sight plane.[1] [2]

Variants

Commando

As WWII began to heat up, the government stepped up procurement of various weapons for the war effort, including a series of orders in 1941-42 for about 5,000 Official Police revolvers destined for defense plant guard duty, security officers of various government bureaus, and Military police use.[6] Already burdened by massive wartime production contracts for the M1911 Colt had difficulty fulfilling these requests, and in addition the procurement agency, the Defense Supplies Corporation (DSC), felt that the $28 per unit pricing was too high.[6] The DSC pressed Colt for a less expensive and simplified version of its commercial model which would hopefully streamline the production process. Colt responded by introducing the "Commando", which was manufactured with either a two or four inch barrel, and non-gloss Parkerized matte blued finish. The Commando also lacked the usual metal checkering on the hammer, trigger, and cylinder latch, as well as the reflection-deadening treatment of the commercial version's top strap. In addition, plastic material replaced the wood grips of the civilian model.[6] Altogether approximately 48,611 Commando war-time models were eventually acquired through mid-1945, most were supplied for sentry and police type functions, although around 12,800 were issued to various intelligence services such as US Military intelligence and the Office of Strategic Services (OSS), usually in the two inch barrel configuration referred to as the "Junior Commando".[6] A few Commandos saw service overseas in the war zone.[6] After hostilities ended, more than 12,000 of the four inch Commandos ended up being bobbed off to two inch models.[6]

Marshall

A rare variant featuring a rounded grip, the Marshall was only produced from 1955-1956. With a very limited production run of 2,500 units, the Marshall became a true collectable.[2]

MK III

The moniker "Official Police" was borrowed by one model in a new generation of revolvers Colt introduced in the late1960s, called the "MK III" series. MK III models consisted of simpler versions of several classic Colt revolvers with updated lockwork.[7] The MK III product line was actually a different and original design based on a new "J" frame, which failed to attain commercial success and was cancelled after only three years.[1] [7]

See also

- .38/200 Enfield Revolver
- S&W Victory Revolver

References

[1] "Colt's Official Police Revolver" (http://www.shootingtimes.com/handgun_reviews/coltp_083106), **Shooting Times** magazine Web site – Handgun Reviews. Accessed August 13, 2008.

[2] Ayoob, Massad. "The Colt Official Police: 61 years of production, 99 years of service" (http://findarticles.com/p/articles/mi_m0BQY/is_5_53/ai_n27176787), **Guns** magazine. BNET Web site – Find articles. Accessed August 13, 2008.

[3] "Colt Official Police" (http://www.bellum.nu/armoury/COfficialPolice.html), Bellum Web site. Accessed August 20, 2008.

[4] Pate, Charles, *U.S. Handguns of World War II*, Andrew Mowbray Publishing, ISBN 100917218752 139780917218750

[5] "Colt Commando" (http://www.bellum.nu/armoury/CCommando.html), Bellum Web site. Accessed August 21, 2008.

[6] "The World War II Commando Revolver" (http://www.manatarmsbooks.com/pate.html), Man at Arms Web site. Accessed August 21, 2008.

[7] "Colt mk. III revolvers: Trooper, Lawman, Official Police (USA)" (http://world.guns.ru/handguns/hg82-e.htm), World.guns.ru Web site. Accessed August 21, 2008.

Colt's Manufacturing Company

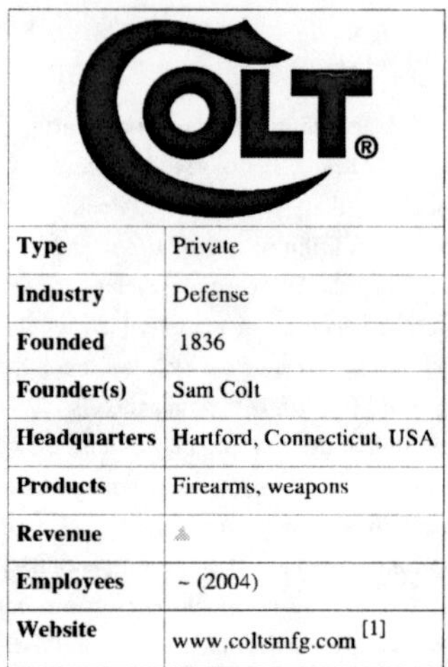

Type	Private
Industry	Defense
Founded	1836
Founder(s)	Sam Colt
Headquarters	Hartford, Connecticut, USA
Products	Firearms, weapons
Revenue	
Employees	~ (2004)
Website	www.coltsmfg.com [1]

Colt's Manufacturing Company (**CMC**, formerly **Colt's Patent Firearms Manufacturing Company**) is a United States firearms manufacturer founded in 1847. It is best known for the engineering, production, and marketing of dozens of different firearms over the later half of the 19th and the 20th century. It has made many civilian and military designs used in the United States, as well was many other countries.

Among the most famous products from Colt are the Walker Colt used by the United States Mounted Rifles in the Mexican-American War and the "Colt .45" revolver, the proper name of which was the Single Action Army or Peacemaker. Later well-known CMC revolvers include the Colt Python and Colt Anaconda. John Browning also worked for Colt for a time, and came up with now ubiquitous parallel slide type of design for a pistol, which debuted on the Colt M1900 pistol, leading to numerous pistol designs including the famous Colt M1911 pistol. Though they did not develop it, Colt was responsible for M16 production for a long time, as well as many derivative firearms related to it. The most successful and famous of these are numerous M16 carbines, including the Colt Commando family, and the M4 carbine.

Colt also developed many important less known firearms that were often ahead of their time. Among the most recent was the CAR-15 family - an innovative weapon system family of the 1960s, as well as a number of 5.56 mm machine guns such as the Colt CMG-1, CMG-2 in the 60s in the 70s. They also invented the Colt SCAMP PDW, a little known firearm of the late 1970s that was among the first of its type. Colt's produced also the first 15 000 Thompson Submachineguns Mod 1921. Another important design was the lesser-known Colt-Browning Model 1895 (Potato Digger) - one of the first gas-actuated machine guns. Going back even farther reveals other important products of the 19th century. The Colt Revolver Rifle, one of the first repeating rifles, and used during the American Civil War. In addition to this were a large number of famous revolvers, such as the 1847 Colt Walker, the smaller Dragoon Mod. 1848 of the same caliber .44, the Navy Mod. 1851 cal .36, the Pocket Mod. 1849 cal .31 and numerous other famous revolvers of the 'Wild West'. His designs played a major role in the popularization of the revolver and the shift away from earlier single pistols and pepperbox type weapons. While Colt did not invent the revolver concept, his designs resulted in the first very successful ones with patents on many of the features that lead to them being so popular.

In 2002, Colt Defense was split off from Colt's Manufacturing Company. Colt Manufacturing Company now serves the civilian market, while Colt Defense serves the law enforcement, military, and private security markets worldwide. Prior to the split Colt was also well known for their production (now taken over by Colt Defense) of the M1911 semi-automatic pistols, M4 carbines, M16 assault rifles, and M203 grenade launchers, although none of these were Colt designs, excepting the M1911 . Diemaco of Canada was also purchased, and renamed Colt Canada, though most of its products remain the same. Diemaco and Colt had earlier worked together on designs and shared many similar products.

History

1836–1911

CMC was founded in Hartford, Connecticut in 1836 by Samuel Colt in order to produce revolvers, of which Colt held the patent, during the Mexican-American War.[2] Colt's earlier venture, the Patent Arms Manufacturing Company, had declared bankruptcy in 1842 and was no longer producing firearms, but the efficiency of the Colt Paterson revolver design had become apparent to the Texas Rangers, and they placed an order for 1,000 larger revolvers that became known as the Walker Colt, ensuring Colt's re-entry into manufacturing revolvers. Later, the U.S. Army also sought out the young entrepreneur to produce even more revolvers.[3]

Colt's early history largely revolved around the production of revolvers, developed out of Sam Colt's original 1834 invention of the revolver.[3] Colt is perhaps best known for the famous "Colt .45", a name which actually refers to four separate historically significant firearms. The first of these is the aforementioned 1873 Single Action Army, of which Colt was the original producer, and which was one of the most prevalent firearms in the American West during the end of the 19th century. Colt still produces this firearm, in six different calibers, two finishes and three barrel lengths. (Original, good condition first generation Single Action Armies, those produced between 1873 and 1941, are among the most valuable to the collector. Especially valuable, often going for well over $10,000, are the Orville W. Ainsworth and the Henry Nettleton inspected U.S. Cavalry Single Action Army Colts.

The second was the Colt Model of 1878. It was Colt's first large frame double action revolver. It combined the front end of the Single Action Army revolver with a double action 6 shot frame mechanism. It was available commercially in numerous calibers including .45. In 1902 the U.S. Army purchased several thousand of the 1878 revolvers in caliber .45 Colt with an over sized trigger guard. These were issued by the US Army to the, then new, Philippine Constabulary Corps. Some of them are also believed to have accompanied US troops to Alaska in the early 1900s. For these reasons the 1902 purchase revolvers are sometimes called the Alaskan or Philippine model. Although long obsolete to the US Army these revolvers were still in use in the Philippines when Japan invaded in 1941.

The third .45 made by Colt was the New Service Double Action revolver. From introduction in 1898 to the beginning of World War II it was a major part of the Colt line. In caliber .45 Colt it was accepted by the U.S. Military as the Model 1909 .45 revolver and replaced the 1873 Single Action revolvers. It primarily differed from the Model 1878 by having a swing out cylinder for faster loading with a new method of ejecting cartridges, and an improved trigger design. The New Service revolver was also available in other calibers such as .38 Special, and later on in the 20th century, .357 Magnum.

One of the first truly modern-style handguns, the Colt revolvers became known as "The Great Equalizer", because they could be loaded and fired by anyone, whereas most previous guns had required sufficient strength and dexterity. In theory, anyone who had a modern-style revolver was equal to anyone else, regardless of their relative physical abilities. This term has since come to be used for firearms in general, as awkward weapons like muzzle-loaded muskets became a thing of the past.

Colt was one of the first companies to create a product with interchangeable parts. At the New York Crystal Palace Exhibition in 1853, a Colt exhibition dissembled ten guns and reassembled ten guns using different parts from different guns.

Though the US was not directly involved in the Crimean War (1854–1856), Colt weapons were used by both sides.

The OWA Colt refers to the earliest issued Single Action Armies which were inspected by Orville W. Ainsworth. O.W. Ainsworth was the ordnance sub-inspector at the Colt factory for approximately the first thirteen months (Oct. 1873 to Nov. 1874) of the Single Action Army's production. It was Ainsworth that inspected the Colts used by General Custer's 7th Cavalry troops at the Battle of the Little Bighorn. However General Custer himself fell holding a couple of English-made Webley revolvers in his hands.

Henry Nettleton was the ordnance inspector in 1878 at the Springfield Armory. Second only to the OWA Colts, Nettleton Colts are prized by serious collectors. Both the Nettleton and OWA Colts will have the cartouche (OWA or HN) on the left side of the wood grip.

The Single Action Army has been copied by numerous makers both in America and in Europe. The two major makers of Colt replicas are Aldo Uberti in Italy and U.S. Fire Arms Mfg. Co. in Hartford, Connecticut.[3]

Under a contract with the U.S. Army Colt Arms built the Model 1895 ten-barrel variant of the Gatling Gun, capable of firing 800-900 .30 Army rounds per minute, and used with great effect at the Battle of San Juan Hill.[4] The M1895 Colt-Browning machine gun or "Potato Digger" was also built by Colt. The Colt-Browning was one of the first gas-operated machine guns, originally invented by by John Browning. It became the first automatic machine gun adopted by the United States and saw limited use by the U.S. Marine Corps at the invasion of Guantánamo Bay and by the 1st Volunteer Infantry in the Santiago campaign during the Spanish-American War.

The Colt entry for a semi-automatic pistol at the turn of the 20th century defeated two other contenders: a .45 Pistol Parabellum (e.g. the Luger pistol) from DWM and an entry from Savage Arms. There had been many other contenders earlier on, but these were eliminated. The Colt also competed with Colt M1900 design in .38 ACP against other entrants in a 1900 competition that included entries from Mauser. The winner evolved into the famous 1911 pistol in 45 ACP, and would be used by the U.S. military for much of the 20th Century and several major wars; variants in 38 Super and other calibers (even 38 Special) and in other barrel lengths found use by civilians and in pistol competition.

1911–1984

Model of 1911 Colt Pistol, U.S. Army, first year of production (1912)

The fourth famous "Colt 45" is the John Browning-designed M1911, which was the standard U.S. military sidearm from 1911 to 1985. The M1911 is still frequently used by civilians, law enforcement, and military agencies today. Variants in other barrel lengths and other calibers (notably 38 Super and 9 mm) have been used extensively in combat shooting and pistol marksmanship, and the guns often are "accurized" into amazingly precise competition tools or custom combat weapons.

During World War I Colt was not able to meet the U.S. military's demand for 1911 production. A decision was then made to accept Colt New Service revolvers in caliber .45 ACP as a substitute weapon. This New Service variant in 45 ACP was called the U.S. Model 1917 revolver. A competing manufacturer, Smith & Wesson, also made double action revolvers in .45 ACP which were also accepted and issued by the U.S. military as Model 1917 revolvers. Extensive use of the 1917 revolvers as well as the 1911 semi-automatic pistol occurred during World War I.

By the end of World War I production of the 1911 had reached high enough numbers so the 1917 revolvers were declared limited standard weapons and large numbers of them were placed in storage until World War II when they again saw service with the US military. Small numbers of the Model 1917 revolvers remained in service within the continental U.S. with the Army's MP Corps until the 1960s.

The period between the world wars was marked by increased adoption of Colt revolvers by many police departments. The most popular police caliber was .38 Special. Colt offered a snub-nosed revolver called the Detective Special, and also marketed a longer barreled small frame .38 as the Police Positive. A heavier, but very similar, revolver on a larger frame was the Colt Official Police. These weapons would dominate the US law enforcement market until the 1950s when Smith and Wesson began to aggressively market their own .38 Special design. During World War II, 2 and 4 inch parkerized versions of the .38 Official Police were purchased by the U.S. military for issuance to Defense Plant guards as the Colt Commando revolver.

Since Auto Ordnance had no tooling for production, Colt acquired the license for the Thompson 1921 SMG and made a first batch of 15,000 pieces the first production year.

Colt's New Service revolver line remained in production until World War II when a decision was made by Colt to move the machinery outside to the parking lot to make room for higher priority military contract production. There it rusted and the production of this model was never resumed. Civilian models of the New Service had been available in many different barrel lengths and calibers. In the 1930s an adjustable sight version called the Shooting Master with checkered grips was marketed.

The end of World War II also saw the end of Colt production of their popular .32 and .380 pocket pistols which had begun in 1903 and 1908 respectively. The U.S. military versions of these pistols was called the Model M. These concealable pistols were purchased by the U.S. military during World War II for usage by couriers, clandestine service operatives, military investigators and were also issued to U.S. Army generals as the General Officer Pistol.

The 1960s were boom years for Colt with the escalation of the Vietnam War, Robert McNamara shutting down the Springfield Armory, and the U.S. Army's subsequent adoption of the M16 (to which Colt held the production rights.)

Colt would capitalize on this with a range of AR-15 derivative carbines. They also developed AR-15 based Squad Automatic Weapons, and the Colt SCAMP, an early PDW design.

The Colt XM148 grenade launcher was created by Colt's Design Project Engineer, gun designer Karl R. Lewis [5]. The May 1967 "Colt's Ink" newsletter announced that he had won a national competition for his selection and treatment of materials in the design. The newsletter stated in part, "In only 47 days, he wrote the specifications, designed the launcher, drew all the original prints, and had a working model built."

At the end of the 1970s, there was a program run by the Air Force, to replace the M1911A1. The Beretta 92S won, but this was contested by the Army. The Army ran their own trials, leading eventually to the Beretta 92F being selected as the M9.

1984–1992

The 1980s marked fairly good years for Colt, but the coming end of the Cold War would change all that. Colt had long left innovation in civilian firearms to their competitors, feeling that the handgun business could survive on their traditional double-action revolver and M1911 designs. Instead, Colt focused on the military market, where they held the primary contracts for production of rifles for the US military.

Colt ACR/M16A2E2 (second from top to bottom), of the U.S. Advanced Combat Rifle program

This strategy dramatically failed for Colt through a series of events in the 1980s. In 1984, the U.S. military standardized on the Beretta 92F. This was not much of a loss for Colt's current business, as M1911A1 production had stopped in 1945, and most had not been made by Colt at the time.

Meanwhile, the military rifle business was growing because the U.S. Military had a major demand for more upgraded M16s, the M16A2 model had just been adopted and the Military needed hundreds of thousands of them.

In 1986, Colt's workers, members of the United Auto Workers went on strike for higher wages. This strike would ultimately last for four years, and was one of the longest running labor strikes in American history. With replacement workers running production, the quality of Colt's firearms began to slip. Dissatisfied with Colt's production, in 1988 the U.S. military awarded the contract for future M16 production to Fabrique Nationale.

Some criticized Colt's range of handgun products in the late 1980s as out of touch with the demands of the market, and their once-vaunted reputation for quality had suffered during the UAW strike. Colt's stable of double action revolvers and single action pistols were seen as old fashioned by a marketplace that was captivated by the new generation of "wondernines" - high-capacity, 9 mm caliber handguns, as typified by the Glock 17.

Realizing that the future of the company was at stake, labor and management agreed to end the strike in an arrangement that resulted in Colt being sold to a group of private investors, the State of Connecticut, and the UAW itself.

The new Colt first attempted to address some of the demands of the market with the production in 1990 of the *Double Eagle*, a double action pistol based heavily on the M1911 design which was seen as an attempt to "modernize" the classic Browning design. Colt followed this up in 1992 with the *Colt All American 2000*, which was unlike any other handgun Colt had produced before.

The Colt All American 2000 was a polymer framed, rotary bolt, 9 mm handgun with a magazine capacity of 15 rounds. It was everything that Colt thought the civilian market wanted in a handgun. Unfortunately, the execution was disastrous. Early models were plagued with inaccuracy and unreliability, and suffered from the poor publicity of having to be recalled. The product launch failed and production of the All American 2000 ended in 1994.

The cost of developing Colt's ACR also cut into their bottom line, as none of the ACR contestants were adopted — a result that came out in the early 1990s.

All of the above ultimately led to the company's chapter 11 bankruptcy in 1992. Colt Manufacturing Co. announced the termination of its production of double action revolvers in October 1999 [6] .

1992–present

The 1990s brought the end of Cold War, which resulted in a large down turn for the entire defense industry. Colt was hit by this downturn, though it would be made worse later in the 1990s by a boycott.

The boycott

In 1994, the assets of Colt were purchased by Zilkha & Co, a financial group owned by Donald Zilkha. It was speculated that Zilkha's financial backing of the company enabled Colt to begin winning back military contracts. In fact during the time period it won only one contract, the M4 carbine. However, the U.S. Military had already been purchasing Colt Carbines for the past 30 years (See Colt Commando).

During a 1998 Washington Post interview, CEO Ron Stewart stated that he would favor a federal permit system with training and testing for gun ownership. This led to a massive grass-roots boycott of Colt's products by gun stores and ordinary gun owners, some of whom sold their Colt firearms to cut into Colt's market share even more. This ultimately led to the resignation of Ron Stewart.

Zilkha replaced Stewart with Steven Sliwa and focused the remainder of Colt's handgun design efforts into "smart guns", a concept which was favored politically but had little interest or support among handgun owners or Police Departments. This research never produced any meaningful results due to the limited technology at the time.

The boycott of Colt has faded out with the new CEO William M. Keys, a retired U.S. Marine Lt. General, working hard to bring Colt back from its tarnished reputation. Due to the efforts of William Keys, Colt's quality has improved as much as its favor with diehard Colt fans.

Increased competition

Most problematic for Colt, its flagship 1911 pistols and AR-15 rifles had to compete with a glut of the company's own used rifles and pistols that could be purchased at prices well below what Colt offered for their new products on the civilian market.

Colt also has to compete with other companies that make 1911-style pistols such as Kimber and AR-15 rifles such as Bushmaster. Bushmaster has subsequently overtaken Colt in the number of AR-15s sold on the civilian market.

Colt suffered a stinging legal defeat in court when it sued Bushmaster for trademark infringement claiming that "M4" was a trademark that it owned. The judge ruled that since the term M4 is a generic designation that Colt does not specifically own, Colt had to pay monetary reimbursement to Bushmaster to recoup Bushmaster's legal fees. The M4 designation itself comes from the U.S. military designation system, whose terms are in the public domain.

Colt continues production of classic designs such as the SAA, sold in both the limited collector's market and through more traditional channels. However, it survives primarily on the manufacturing of a variety of civilian and military weapons. The most popular of these are various AR-15 Carbines, a weapon category that it invented and helped develop over nearly 30 years since acquiring the AR-15 design. The AR-15 Carbine derivatives, and weapons like them have proved so popular that a large amount of competition has arisen in the area. As with AR-15 rifles, the original Colt designs and their derivatives are heavily copied, and as a result they face much competition from other manufacturers, including Springfield Armory, Kimber Manufacturing, Rock River Arms, and US Fire Arms.

Colt has entered in several US contracts with mixed results. For example, Colt had an entry in the Advanced Combat Rifle (ACR) program of the 1980s, but along with other contestants failed to replace the M16A2. Colt and many other makers entered the US trials for a new pistol in the 1980s, though the Beretta entry would win and become the M9 Pistol. The Colt OHWS handgun was beaten by H&K for what became the MK23 SOCOM, it was lighter than the H&K entry but lost in performance. Colt did not get to compete for the XM8 since it was not an open competition. Colt is a likely entrant in any competition for a new US service rifle. Current M16 rifles have been made primarily by FN USA since 1988. However, Colt remains the sole source for M4 carbines for the US military. Under their license agreement with Colt, the US military cannot legally award second-source production contracts for the M4 until July 1, 2009.

Firearms

Selected famous or innovative Colt products

Handguns

The years in brackets indicate the year when production started, not the year of the model's patent.

- Colt Paterson revolver (1836)
- Walker Colt revolver (1847)
- Colt Dragoon revolver (1848)
- Colt 1851 Navy revolver (1851)
- Colt Army revolver (1860)
- Colt 1861 Navy revolver (1861)
- Colt Model 1862 revolver (1862)
- Colt Single Action Army "Peacemaker" revolver (1873)
- Colt Lightning, Thunderer and Rainmaker revolvers (1877)
- Colt M1894 Army DA Revolver (1894)
- Colt M1900 semiautomatic pistol (1900)
- Colt Model 1903 Pocket Hammerless semiautomatic pistol (Model M, 1903)
- Colt Model 1908 Vest Pocket (Model N, 1908)
- Colt M1911 semiautomatic pistol; also known as the Government Model (Model O, 1911)
- Colt M1917 revolver / New Service / M1909 / Colt Shooting Master
- Colt 2000 semi-automatic 9 mm
- Colt Anaconda revolver (AA Frame)
- Colt Cadet 22
- Colt Delta Elite a modified Colt M1911A1 chambered for the 10mm Auto
- Colt Detective Special / Cobra / Agent / snubnosed revolver (D frame)
- Colt Diamondback revolver
- Colt Double Eagle Pistol
- Colt Grizzly revolver
- Colt Junior
- Colt King Cobra revolver
- Colt New Agent semiautomatic pistol
- Colt Official Police revolver / Officers Model (Match, Target & Special) / New Army & Navy (E/I frame) / Colt .357 (I Frame)
- Colt Police Positive revolver
- Colt Police Positive Special revolver / Viper (D Frame)
- Colt SCAMP
- Colt SF VI, DS II, Magnum Carry (.357 Magnum)
- Colt Python revolver (I frame)
- Colt Trooper revolver (I frame), Trooper Mk III (J frame), and trooper Mk V (V Frame)
- Woodsman/Woodsman Match Target/Huntsman/Targetsman (Model S)

Colt Diamondback .22

Long guns

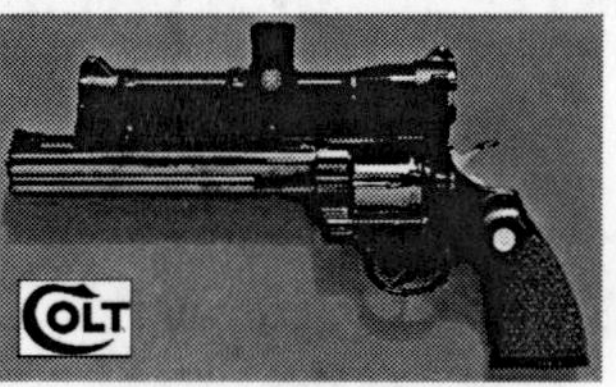

Colt Python Silhouette .357 Magnum

- Colt-Browning M1895 machine gun
- Colt Lightning Carbine
- Colt 1861 Special Contract Musket
- AR-15 type rifles, such as the M16 rifle, M4 carbine and Colt Commando (*see AR-15 variants for a complete list*)
- Colt ACR
- Various Shotguns

Colt also manufactured several military long arms under contract including the M1918 BAR and Thompson SMG.

Colt Anaconda .44 Magnum

See also

- List of modern armament manufacturers
- List of firearms
- Colt Canada
- Antique Guns
- Sodium silicate (used as a cement for paper cartridges used in early Colt revolvers during the American Civil War)

External links

- Colt Defense [7]
- Colt's Manufacturing Company [1]
- Colt Automatic Pistols Home Page [8]
- Bushmaster wins trademark case brought by Colt [9]
- "United States Utility Patent 1304, Improvement in fire-arms and in the apparatus used therewith" [10]. United States Patent Office; Google. Retrieved 2008-09-02.

References

[1] http://www.coltsmfg.com/

[2] Flayderman, Norm (2001). *Flayderman's Guide to Antique American Firearms... and their values*. Iola, WI: Krause Publications. p. 669. ISBN 0-87349-313-3.

[3] Smith, WHB (1968). *Book of Pistols and Revolvers*. Harrisburg, PA: Stackpole Books.

[4] Parker, John H. (Lt.), *History of the Gatling Gun Detachment*, Kansas City, MO: Hudson-Kimberly Publishing Co. (1898), pp. 131-138

[5] http://www.rareguncollection.com/

[6] http://calnra.com/COLTupdate.shtml

[7] http://www.colt.com/

[8] http://www.coltautos.com/

[9] http://www.bushmaster.com/bushmaster_wins_trademark_case.asp

[10] http://www.google.com/patents?id=bNI_AAAAEBAJ&dq=Samuel+Colt

Article Sources and Contributors

Enfield revolver *Source*: http://en.wikipedia.org/w/index.php?oldid=379687691 *Contributors*: 171046, 5infBrig, Andrwsc, Asams10, Atirador, Bobblewik, Brighterorange, Commander Zulu, DJ Tricky86, DocWatson42, Dumelow, DutchDevil, Edward, GPS73, GamerJay, GraemeLeggett, GregorB, Ilion2, Iridescent, JKBrooks85, Jaksmata, Kbthompson, Koalorka, Koavf, Kozuch, Kross, Lightmouse, Mattbr, MatthewVanitas, Mehrunes Dagon, Monarchist5, Morphinea, Nemo5576, Nukes4Tots, Paul Broadway, Pyrotec, Quebec99, Rama, Resist, Rettetast, Rjwilmsi, Ryanrs, Saga City, Silent SAM, SirIsaacBrock, Skrunyak, Surv1v4l1st, Thaurisil, Thernlund, Thewumpus, Tony1, Trekphiler, Tubezone, Veritas Panther, 67 anonymous edits

Royal Small Arms Factory *Source*: http://en.wikipedia.org/w/index.php?oldid=374231749 *Contributors*: Alansplodge, Aldis90, Chrism, Commander Zulu, Czyrko, Dormskirk, Dvavasour, Flosssock, GraemeLeggett, Greenshed, Gurch, Hmains, Hugo999, Isaac Sanolnacov, JackyR, Jaraalbe, Johnlp, Kartano, Kbthompson, Kevin Murray, Longhair, Lupin, Mark83, Monsted, Morphinea, Nono64, Northmetpit, Postlebury, Psarj, Pyrotec, Rayxt, Rjwilmsi, Saburny, Saga City, SimonP, That Guy, From That Show!, TheWatcherREME, Thernlund, Una Smith, WOSlinker, Yadayadayaday, Yaf, 15 anonymous edits

Service pistol *Source*: http://en.wikipedia.org/w/index.php?oldid=379299995 *Contributors*: Akyoyo94, Aldis90, AllStarZ, Alucard365, Aptospalma831, Asams10, Beanerschnitzel, Becritical, Beşiktaşlı48, Boris Barowski, CSWarren, Carnildo, Circeus, Closedmouth, Commander Zulu, CommonsDelinker, Deathbunny, Deon Steyn, Doc Dish, Download, Ennerk, Garion96, Jack Bethune, Jellyfish dave, Jetwave Dave, Joffeloff, Josh Woolstenhulme, Kimchi.sg, Kintetsubuffalo, Koalorka, Krj373, Kross, Kwiki, Lightmouse, MBK004, MatthewVanitas, Mrg3105, Museerouge, Nick-D, Olegvolk, Phil1988, Pinkadelica, Puddhe, Quickload, R'n'B, Rama, Rockz, Scoo, Septegram, Skysmith, Steven Luo, Sus scrofa, TaintedMustard, TheTranc, Thernlund, ThreeBlindMice, Thumperward, Vegaswikian, Vikingviolinist, Woohookitty, WotWeiller, Xezbeth, Товарищ, 186 anonymous edits

Webley Revolver *Source*: http://en.wikipedia.org/w/index.php?oldid=379689821 *Contributors*: 4u1e, 5infBrig, Acad Ronin, Accurizer, Aksi great, Aldis90, AliveFreeHappy, Alyoshenka, Andplus, Angusmclellan, Asams10, Asatruer, Ashley Pomeroy, AtTheAbyss, Atirador, AuburnPilot, Bailitothemexicanfajita, Bigblackdilldo, BlueOrb, Bobblewik, Boreas74, Brad Lawrence, Brumburger, Carguychris, Commander Zulu, Corporal Punishment, Cqms, DRAC250, Dakarii2, DanielCD, Danny1232005, Darkman IV, David Underdown, Dawkeye, Dellxps, Deon Steyn, Dincher, Doc glasgow, DocWatson42, DragonflySixtyseven, Dreamafter, Droll, Emediarifleman, EnigmaMcmxc, F, Fatfuker, Finster72, Francis45, Fsdajkl, GPS73, Geniac, Geoff B, GoldDragon, GraemeLeggett, Groovenstein, HDCase, Hagger, Happysailor, Hidoshi, Hlvirton, Hmaag, Hmains, Hut 8.5, Ian Dunster, JSHS09, Jackas112, James Anthony Knight, Jhliggett, Jons63, Kaeso Dio, Kaiserb, Kbdank71, Kernel Saunters, Kimchi.sg, Kirill Lokshin, Koalorka, Kozuch, Kross, Kubigula, Kukini, LWF, Lankiveil, Lemmey, Liamdowney, Lightmouse, Lights, Macgyver-bd 896, Malo, Maralia, MarcK, Mattbr, MatthewVanitas, Mdleon8863, MegX, Mervyn, Michael Devore, Michael Dorosh, Mickyjayohyeah, Morphinea, Moviemaniacx, Naaman Brown, Nabokov, Narson, NuclearWarfare, Nukes4Tots, Oberiko, Ortolan88, Oxymoron83, PDH, Particle man, Peripitus, PiCo, Pigsonthewing, Piratedan, Pooppooppooppoop, ROG5728, Rama, Raul654, Redxiv, Rettetast, Riddley, Rifleman 82, Ritwikbmca, Rusfuture, SallyForth123, SandyGeorgia, Sigint74, Signalhead, SirIsaacBrock, Skrunyak, Some guy, Spellmaster, SqueakBox, Suruena, Surv1v4l1st, Swatjester, TKD, Technopat, The Mystery Man, Theredstarswl, Thernlund, Thomphson, Tiptoety, TnoWatanabe, Tony1, Trekphiler, Twinxor, Ultimablah, Ultrabeater, Vanwall, Ve3, Veritas Panther, Voloshinov, Weebenginger, West Brom 4ever, Wikinick, Winged Brick, Wknight94, Woohookitty, Your username goes here, ZulaFritz, 211 anonymous edits

Smith & Wesson Victory Model *Source*: http://en.wikipedia.org/w/index.php?oldid=232896019 *Contributors*: 5infBrig, Accurizer, Aldis90, Alexius08, Asams10, Bjuice, Bloodshedder, Commander Zulu, CommonsDelinker, Coyote5150, Deon Steyn, Fat pig73, Geeman470, GlassCobra, Hmains, Inwind, JPG-GR, Jack Bethune, Jetwave Dave, Knollbert, Koalorka, LWF, Leisuresuit, Lord Bodak, MOOOOOPS, Major tom, MatthewVanitas, Mcumpston, Mendaliv, Mike Searson, MikePGS, Mvialt, Nabokov, Nef, New Hampshirite, Nukes4Tots, Olegvolk, Phil1988, Police,Mad,Jack, Quest for Truth, Reikon, Rubensni, SAWGunner89, SirIsaacBrock, Sukiari, Surv1v4l1st, The Cake is a Lie, The searcher, Thernlund, Thewellman, Thunderbuster, Timothylord, Ulric1313, Vegaswikian, Vicpol, Yhinz17, ZH Evers, 86 anonymous edits

Colt New Service *Source*: http://en.wikipedia.org/w/index.php?oldid=371426473 *Contributors*: Aldis90, BobbieCharlton, Commander Zulu, GrantRCanada, Ground Zero, HarryPagetFlashman, Hmains, Ilion2, Jswikart, KEN, MatthewVanitas, MrMontag, Nukes4Tots, Trekphiler, Wmoberndorf, 8 anonymous edits

Smith & Wesson No. 3 Revolver *Source*: http://en.wikipedia.org/w/index.php?oldid=375401725 *Contributors*: Aldis90, AliveFreeHappy, Atirador, Boris Barowski, Chris the speller, Commander Zulu, DOHC Holiday, DocWatson42, Ettrig, Hmaag, Inwind, Keserman, MatthewVanitas, Nemo5576, Nightkey, PigFlu Oink, Riddley, S, Thernlund, Vegaswikian, 13 anonymous edits

Double action only *Source*: http://en.wikipedia.org/w/index.php?oldid=82470069 *Contributors*: A Nobody, A8UDI, ABF, Akyoyo94, AliveFreeHappy, Altenmann, Alucard365, Amabadboy, Anonymous editor, Anthony Appleyard, Asams10, Bart133, Blinkythecrow, Bobbfwed, Bongwarrior, Cannibalicious!, Captpackrat, Colt1911a1s, Cometstyles, Commander Keane, Commander Zulu, CommonsDelinker, DanMP5, Dbooksta, Deathbunny, Doradus, Dorftrottel, Drjava2001, EEMIV, Eastlaw, Eekerz, Emote, Francis Flinch, GFellows, Hatchetfish, Hooperbloob, Howa0082, Hyram.Y.Xu, Insanity Incarnate, Jaw1964, LWF, Lbmcse, Liko81, MCTales, Marqueed, Mavin 101, Mordicai, NAHID, Nakon, Night Gyr, NineseveN, Nukes4Tots, Ocn169, Patrick, Per Honor et Gloria, Peyre, Pgk, Phaerus, Phantomsteve, Rackham, RelentlessRecusant, Rhombus, Riddley, Robomaeyhem, Santaman223, ScAvenger, Sceptre, Seano1, Sevesteen, Sladuuch, Straker, SunCreator, TheHerbalGerbil, TheTito, Thernlund, Tmaull, Tompot, Trevjos, Vicarious, W, Winged Brick, Wizard191, Xanzzibar, Yaf, 149 anonymous edits

Colt Official Police *Source*: http://en.wikipedia.org/w/index.php?oldid=344348097 *Contributors*: Aldis90, Fratrep, Hmains, MatthewVanitas, New Hampshirite, Rama, TDogg310, Wifione, Wikidenizen, 15 anonymous edits

Colt's Manufacturing Company *Source*: http://en.wikipedia.org/w/index.php?oldid=379507845 *Contributors*: A. Dupin, AdamBMorgan, Ahoerstemeier, Alai, Alex43223, AliveFreeHappy, Asams10, Ashwin18, Austin Hair, BTDLBOY, Beginning, Betacommand, Biolith, Boris Barowski, Boycew39, Can't sleep, clown will eat me, Ccunning, Cholmes75, Chris the speller, Chuckstar, Clokemg, Clsykora, Cremepuff222, CrucifiedChrist, CrypticBacon, Curiosandrelics, D.E. Watters, DMCer, Dachannien, DanMP5, Deathbunny, Dellant, Deon Steyn, Dictouray, Dispenser, DocWatson42, DuaneThomas, Edward321, Elipongo, Ellywa, Emediarifleman, Enviroboy, Eranb, Fg2, Francis Flinch, GDO2010, Gaius Cornelius, Gliu, GraemeL, Grant65, Greenshed, Greenwave75, GregorB, Gryffindor, Guilty@eskimo.com, Hayden120, Hede2000, Hephaestos, Hmaag, Hoplon, Hotspur23, HowardSelsam, I already forgot, J rath, J.delanoy, Jeff dean, JodyB, Jrkarp, JudithSouth, Kazvorpal, Kimber1911, Kintaro, Kkesler, KnowledgeOfSelf, Kross, LWF, LindaWarheads, LtPowers, Ltvine, Luna Santin, MBK004, MapleTree, Maralex334, Marshall Stax, MatthewVanitas, Mboverload, Michael Hardy, Mike Searson, Monkeyleg, Motorrad-67, Murf145, Newportm, Nightscream, Nukes4Tots, O^O, Ominae, Paul J Williams, Pearle, PianoKeys, Pibwl, Pierre cb, Plastic Fish, Puchiko, Quickload, R69S, Riddley, Roger Gianni, Roundeyesamurai, SU Linguist, SaltyBoatr, Sam Hocevar, Scott Paeth, ScottyBoy900Q, Search4Lancer, SireMarshall, Special-T, Spitfire19, Stupid girl, Sukiari, Tabletop, Tbrittreid, Tea King, The Anome, TheNightRyder, Themfromspace, Thernlund, Tlesher, Tmorton166, Trasel, Tronno, Tsange, Ugen64, Vargklo, Ve3, Vegaswikian, Walrus19, Wikidenizen, Wikster E, William Allen Simpson, Xeworlebi, Xnatedawgx, Yaf, YoungFreud, Zeusdog, Zzyzx11, 237 anonymous edits

Image Sources, Licenses and Contributors

Image:Enfield Mk II revolver.JPG *Source*: http://en.wikipedia.org/w/index.php?title=File:Enfield_Mk_II_revolver.JPG *License*: Attribution *Contributors*: Rama

File:Flag of the United Kingdom.svg *Source*: http://en.wikipedia.org/w/index.php?title=File:Flag_of_the_United_Kingdom.svg *License*: Public Domain *Contributors*: User:Zscout370

Image:Revolver Enfield No2 Mk I.jpg *Source*: http://en.wikipedia.org/w/index.php?title=File:Revolver_Enfield_No2_Mk_I.jpg *License*: Attribution *Contributors*: Herrick, Nemo5576, Rama

Image:Webley Military Mark IV 1793.jpg *Source*: http://en.wikipedia.org/w/index.php?title=File:Webley_Military_Mark_IV_1793.jpg *License*: Attribution *Contributors*: Nemo5576, Sandpiper

Image:Enfield-No2.jpg *Source*: http://en.wikipedia.org/w/index.php?title=File:Enfield-No2.jpg *License*: Creative Commons Attribution-Sharealike 2.0 *Contributors*: User:Rama

File:Flag of Australia.svg *Source*: http://en.wikipedia.org/w/index.php?title=File:Flag_of_Australia.svg *License*: Public Domain *Contributors*: Ian Fieggen

File:Flag of Canada.svg *Source*: http://en.wikipedia.org/w/index.php?title=File:Flag_of_Canada.svg *License*: Public Domain *Contributors*: User:E Pluribus Anthony, User:Mzajac

File:Flag of The Gambia.svg *Source*: http://en.wikipedia.org/w/index.php?title=File:Flag_of_The_Gambia.svg *License*: Public Domain *Contributors*: Atamari, Avala, Denniss, Fry1989, Klemen Kocjancic, Mattes, Neq00, Nightstallion, Porao, ThomasPusch, Vzb83, WikipediaMaster, Zscout370, 3 anonymous edits

File:Flag of Lesotho.svg *Source*: http://en.wikipedia.org/w/index.php?title=File:Flag_of_Lesotho.svg *License*: Creative Commons Attribution-Sharealike 2.5 *Contributors*: User:Zscout370

Image:Royal Small Arms Factory.jpg *Source*: http://en.wikipedia.org/w/index.php?title=File:Royal_Small_Arms_Factory.jpg *License*: unknown *Contributors*: Christine Matthews

Image:L1A1 DM-ST-91-12005.jpg *Source*: http://en.wikipedia.org/w/index.php?title=File:L1A1_DM-ST-91-12005.jpg *License*: Public Domain *Contributors*: SSGT. J.R. RUARK

File:Royal Small Arms Factory1.JPG *Source*: http://en.wikipedia.org/w/index.php?title=File:Royal_Small_Arms_Factory1.JPG *License*: Public Domain *Contributors*: Northmetpit

file:USSROfficerTT33.JPG *Source*: http://en.wikipedia.org/w/index.php?title=File:USSROfficerTT33.JPG *License*: Public Domain *Contributors*: Alex Bakharev, Benlisquare, Commander Zulu, Dcoetzee, Justass, Nemo5576, 3 anonymous edits

file:Mannlicher M1905 AdamsGuns.jpg *Source*: http://en.wikipedia.org/w/index.php?title=File:Mannlicher_M1905_AdamsGuns.jpg *License*: Attribution *Contributors*: Nemo5576

file:MWP Revolvers GPUprising.JPG *Source*: http://en.wikipedia.org/w/index.php?title=File:MWP_Revolvers_GPUprising.JPG *License*: GNU Free Documentation License *Contributors*: Halibutt

file:Mauser C96 AdamsGuns.jpg *Source*: http://en.wikipedia.org/w/index.php?title=File:Mauser_C96_AdamsGuns.jpg *License*: Attribution *Contributors*: Nemo5576, 1 anonymous edits

file:Lahti L-35-1.jpg *Source*: http://en.wikipedia.org/w/index.php?title=File:Lahti_L-35-1.jpg *License*: Creative Commons Attribution-Sharealike 3.0 *Contributors*: M62

file:MAS Mle 1892 1685.jpg *Source*: http://en.wikipedia.org/w/index.php?title=File:MAS_Mle_1892_1685.jpg *License*: Attribution *Contributors*: KTo288, Nemo5576, Zil

file:DWM 4 inch Navy Luger 859.jpg *Source*: http://en.wikipedia.org/w/index.php?title=File:DWM_4_inch_Navy_Luger_859.jpg *License*: Attribution *Contributors*: Ken, Nemo5576, 3 anonymous edits

file:Frommerstop.jpg *Source*: http://en.wikipedia.org/w/index.php?title=File:Frommerstop.jpg *License*: Creative Commons Attribution-Sharealike 2.5 *Contributors*: Beao, Red, UnneededAplomb, 2 anonymous edits

file:Beretta Model 1934 Pistol.jpg *Source*: http://en.wikipedia.org/w/index.php?title=File:Beretta_Model_1934_Pistol.jpg *License*: Attribution *Contributors*: AuburnPilot, M62, Nemo5576

file:Nambupistol2465.jpg *Source*: http://en.wikipedia.org/w/index.php?title=File:Nambupistol2465.jpg *License*: Creative Commons Attribution-Sharealike 3.0 *Contributors*: Oleg Volk Original uploader was Olegvolk at en.wikipedia

file:Pistol TT33.jpg *Source*: http://en.wikipedia.org/w/index.php?title=File:Pistol_TT33.jpg *License*: Attribution *Contributors*: Nemo5576

file:Husqvarna m1907 1777.jpg *Source*: http://en.wikipedia.org/w/index.php?title=File:Husqvarna_m1907_1777.jpg *License*: Attribution *Contributors*: Nemo5576, Piotrek91

file:Swiss Model 1882 Ordnance Revolver 1763.jpg *Source*: http://en.wikipedia.org/w/index.php?title=File:Swiss_Model_1882_Ordnance_Revolver_1763.jpg *License*: Attribution *Contributors*: Amada44, Nemo5576, Rama

file:Webley-Mk-IV-p1030100.jpg *Source*: http://en.wikipedia.org/w/index.php?title=File:Webley-Mk-IV-p1030100.jpg *License*: Creative Commons Attribution-Sharealike 2.0 *Contributors*: User:Rama

file:Webley MkI P0.jpg *Source*: http://en.wikipedia.org/w/index.php?title=File:Webley_MkI_P0.jpg *License*: Public Domain *Contributors*: User:Rusfuture

file:M1911A1 and M9 DA-SC-91-10188.jpg *Source*: http://en.wikipedia.org/w/index.php?title=File:M1911A1_and_M9_DA-SC-91-10188.jpg *License*: Public Domain *Contributors*: AuburnPilot, Jarekt, Nemo5576, Signaleer, 1 anonymous edits

Image:Webley-Mk-IV-p1030100.jpg *Source*: http://en.wikipedia.org/w/index.php?title=File:Webley-Mk-IV-p1030100.jpg *License*: Creative Commons Attribution-Sharealike 2.0 *Contributors*: User:Rama

File:Webley_MkI_P0.jpg *Source*: http://en.wikipedia.org/w/index.php?title=File:Webley_MkI_P0.jpg *License*: Public Domain *Contributors*: User:Rusfuture

Image:Webley-Mark-IV-p1030061.jpg *Source*: http://en.wikipedia.org/w/index.php?title=File:Webley-Mark-IV-p1030061.jpg *License*: Creative Commons Attribution-Sharealike 2.0 *Contributors*: User:Rama

Image:Webley-Mk-IV-p1030102.jpg *Source*: http://en.wikipedia.org/w/index.php?title=File:Webley-Mk-IV-p1030102.jpg *License*: Creative Commons Attribution-Sharealike 2.0 *Contributors*: User:Rama

File:Webley 1868 RIC.JPG *Source*: http://en.wikipedia.org/w/index.php?title=File:Webley_1868_RIC.JPG *License*: Creative Commons Attribution-Sharealike 3.0 *Contributors*: User:Hmaag

File:Webley W. G. Model.JPG *Source*: http://en.wikipedia.org/w/index.php?title=File:Webley_W._G._Model.JPG *License*: Creative Commons Attribution-Sharealike 3.0 *Contributors*: User:Hmaag

Image:455in SAA Ball - Webley 455 Ammunition.jpg *Source*: http://en.wikipedia.org/w/index.php?title=File:455in_SAA_Ball_-_Webley_455_Ammunition.jpg *License*: Creative Commons Attribution-Sharealike 2.5 *Contributors*: Ian Dunster, Malis

File:Webley MkI P0.jpg *Source*: http://en.wikipedia.org/w/index.php?title=File:Webley_MkI_P0.jpg *License*: Public Domain *Contributors*: User:Rusfuture

Image:380RevolverMkIIz Cartridges.JPG *Source*: http://en.wikipedia.org/w/index.php?title=File:380RevolverMkIIz_Cartridges.JPG *License*: Attribution *Contributors*: Original uploader was Commander Zulu at en.wikipedia

File:IOF-32-REV-1.JPG *Source*: http://en.wikipedia.org/w/index.php?title=File:IOF-32-REV-1.JPG *License*: Public Domain *Contributors*: Anupam Kamal

Image:WebleyBulldogReplica.JPG *Source*: http://en.wikipedia.org/w/index.php?title=File:WebleyBulldogReplica.JPG *License*: Attribution *Contributors*: user:Commander Zulu

Image:Webleypocket.jpg *Source*: http://en.wikipedia.org/w/index.php?title=File:Webleypocket.jpg *License*: Creative Commons Attribution-Sharealike 3.0 *Contributors*: User:MatthewVanitas

Image:M&Prevolver.jpg *Source*: http://en.wikipedia.org/w/index.php?title=File:M&Prevolver.jpg *License*: Creative Commons Attribution 2.5 *Contributors*: Original uploader was Olegvolk at en.wikipedia

File:Flag of the United States.svg *Source*: http://en.wikipedia.org/w/index.php?title=File:Flag_of_the_United_States.svg *License*: Public Domain *Contributors*: User:Dbenbenn, User:Indolences, User:Jacobolus, User:Technion, User:Zscout370

Image:M&P1899.JPG *Source*: http://en.wikipedia.org/w/index.php?title=File:M&P1899.JPG *License*: Public Domain *Contributors*: User:Mcumpston

Image:S&W Model 10 original lockwork.jpg *Source*: http://en.wikipedia.org/w/index.php?title=File:S&W_Model_10_original_lockwork.jpg *License*: Public Domain *Contributors*: User:Mcumpston

Image:Fifthchangelock.jpg *Source*: http://en.wikipedia.org/w/index.php?title=File:Fifthchangelock.jpg *License*: Public Domain *Contributors*: Mike Cumpston

File:Flag of Austria.svg *Source*: http://en.wikipedia.org/w/index.php?title=File:Flag_of_Austria.svg *License*: Public Domain *Contributors*: User:SKopp

File:Flag of Hong Kong.svg *Source*: http://en.wikipedia.org/w/index.php?title=File:Flag_of_Hong_Kong.svg *License*: Public Domain *Contributors*: Artem Karimov, ButterStick, Dbenbenn, Homo lupus, Jackl, Kibinsky, Kookaburra, Ludger1961, Mattes, Neq00, Nightstallion, Runningfridgesrule, Sangjinhwa, Shinjiman, ThomasPusch, VAIO HK, Vzb83, Zscout370, 7 anonymous edits

File:Flag of Ireland.svg *Source*: http://en.wikipedia.org/w/index.php?title=File:Flag_of_Ireland.svg *License*: Public Domain *Contributors*: User:SKopp

File:Flag of Israel.svg *Source*: http://en.wikipedia.org/w/index.php?title=File:Flag_of_Israel.svg *License*: Public Domain *Contributors*: AnonMoos, Bastique, Bobika, Brown spite, Captain Zizi, Cerveaugenie, Drork, Etams, Fred J, Fry1989, Himasaram, Homo lupus, Humus sapiens, Klemen Kocjancic, Kookaburra, Madden, Neq00, NielsF, Nightstallion, Oren neu dag, Patstuart, PeeJay2K3, Pumbaa80, Ramiy, Reisio, SKopp, Technion, Typhix, Valentinian, Yellow up, Zscout370, 31 anonymous edits

File:Flag of Macau.svg *Source*: http://en.wikipedia.org/w/index.php?title=File:Flag_of_Macau.svg *License*: unknown *Contributors*: User:PhiLiP

File:Flag of Malaysia.svg *Source*: http://en.wikipedia.org/w/index.php?title=File:Flag_of_Malaysia.svg *License*: Public Domain *Contributors*: User:SKopp

File:Flag of New Zealand.svg *Source*: http://en.wikipedia.org/w/index.php?title=File:Flag_of_New_Zealand.svg *License*: Public Domain *Contributors*: Adambro, Arria Belli, Bawolff, Bjankuloski06en, ButterStick, Denelson83, Donk, Duduziq, EugeneZelenko, Fred J, Fry1989, Hugh Jass, Ibagli, Jusjih, Klemen Kocjancic, Mamndassan, Mattes, Nightstallion, O, Peeperman, Poromiami, Reisio, Rfc1394, Shizhao, Tabasco, Transparent Blue, Väsk, Xufanc, Zscout370, 35 anonymous edits

File:Flag of Norway.svg *Source*: http://en.wikipedia.org/w/index.php?title=File:Flag_of_Norway.svg *License*: Public Domain *Contributors*: User:Dbenbenn

File:Flag of Pakistan.svg *Source*: http://en.wikipedia.org/w/index.php?title=File:Flag_of_Pakistan.svg *License*: Public Domain *Contributors*: Abaezriv, AnonMoos, Badseed, Dbenbenn, Duduziq, Fry1989, Gabbe, Himasaram, Homo lupus, Juiced lemon, Klemen Kocjancic, Mattes, Neq00, Pumbaa80, Rfc1394, Srtxg, ThomasPusch, Túrelio, Zscout370, 7 anonymous edits

File:Flag of Peru.svg *Source*: http://en.wikipedia.org/w/index.php?title=File:Flag_of_Peru.svg *License*: Public Domain *Contributors*: User:Dbenbenn

File:Flag of South Africa.svg *Source*: http://en.wikipedia.org/w/index.php?title=File:Flag_of_South_Africa.svg *License*: Public Domain *Contributors*: User:SKopp

Image:Colt New Service 1370.jpg *Source*: http://en.wikipedia.org/w/index.php?title=File:Colt_New_Service_1370.jpg *License*: Attribution *Contributors*: Huggorm, MichaelMaggs, Nemo5576, 2 anonymous edits

Image:Smith & Wesson No. 3 Third Model Russian 867.jpg *Source*: http://en.wikipedia.org/w/index.php?title=File:Smith_&_Wesson_No._3_Third_Model_Russian_867.jpg *License*: Attribution *Contributors*: Multichill, Nemo5576

Image:S&W New Nodel 3 Frontier 1791.jpg *Source*: http://en.wikipedia.org/w/index.php?title=File:S&W_New_Nodel_3_Frontier_1791.jpg *License*: Attribution *Contributors*: Avron, Kilom691, Model3SW, Nemo5576, 3 anonymous edits

Image:Trigger mechanism bf 1923.jpg *Source*: http://en.wikipedia.org/w/index.php?title=File:Trigger_mechanism_bf_1923.jpg *License*: Public Domain *Contributors*: Unknown (Collective work)

File:Colt Official Police 32-20 1927.png *Source*: http://en.wikipedia.org/w/index.php?title=File:Colt_Official_Police_32-20_1927.png *License*: Creative Commons Attribution-Sharealike 3.0 *Contributors*: User:Hatchetfish

Image:Colt logo.svg *Source*: http://en.wikipedia.org/w/index.php?title=File:Colt_logo.svg *License*: unknown *Contributors*: Vargklo, Wknight94, 1 anonymous edits

Image:Model of 1911 US Army.jpg *Source*: http://en.wikipedia.org/w/index.php?title=File:Model_of_1911_US_Army.jpg *License*: Creative Commons Attribution-Sharealike 3.0 *Contributors*: User:Curiosandrelics

Image:Colt ACR.jpg *Source*: http://en.wikipedia.org/w/index.php?title=File:Colt_ACR.jpg *License*: Public Domain *Contributors*: Nemo5576, Sanandros, Skeezix1000, Thatguy96, 3 anonymous edits

Image:Colt-diamondback.jpg *Source*: http://en.wikipedia.org/w/index.php?title=File:Colt-diamondback.jpg *License*: Attribution *Contributors*: Original uploader was Jeff dean at en.wikipedia

Image:Python-silhouette.jpg *Source*: http://en.wikipedia.org/w/index.php?title=File:Python-silhouette.jpg *License*: Creative Commons Attribution-Sharealike 2.5 *Contributors*: Original uploader was Jeff dean at en.wikipedia

Image:Anaconda-500.jpg *Source*: http://en.wikipedia.org/w/index.php?title=File:Anaconda-500.jpg *License*: Creative Commons Attribution-Sharealike 2.5 *Contributors*: Original uploader was Jeff dean at en.wikipedia

License

CPSIA information can be obtained at www.ICGtesting.com
Printed in the USA
LVOW050758260612

287678LV00003B/6/P